The Rise and Fall of Imperial Chemical Industries

Esther Leslie

The Rise and Fall of Imperial Chemical Industries

Synthetics, Sensism and the Environment

Esther Leslie
School of Creative Arts,
Culture and Communication
Birkbeck, University of London
London, UK

ISBN 978-3-031-37431-9   ISBN 978-3-031-37432-6 (eBook)
https://doi.org/10.1007/978-3-031-37432-6

© The Author(s), under exclusive license to Springer Nature Switzerland AG 2023

This work is subject to copyright. All rights are solely and exclusively licensed by the Publisher, whether the whole or part of the material is concerned, specifically the rights of translation, reprinting, reuse of illustrations, recitation, broadcasting, reproduction on microfilms or in any other physical way, and transmission or information storage and retrieval, electronic adaptation, computer software, or by similar or dissimilar methodology now known or hereafter developed.
The use of general descriptive names, registered names, trademarks, service marks, etc. in this publication does not imply, even in the absence of a specific statement, that such names are exempt from the relevant protective laws and regulations and therefore free for general use.
The publisher, the authors, and the editors are safe to assume that the advice and information in this book are believed to be true and accurate at the date of publication. Neither the publisher nor the authors or the editors give a warranty, expressed or implied, with respect to the material contained herein or for any errors or omissions that may have been made. The publisher remains neutral with regard to jurisdictional claims in published maps and institutional affiliations.

Cover illustration: © Melisa Hasan

This Palgrave Macmillan imprint is published by the registered company Springer Nature Switzerland AG
The registered company address is: Gewerbestrasse 11, 6330 Cham, Switzerland

Acknowledgements

Thanks to the archivists at the library of the Institute of Mechanical Engineers, London; the Wellcome Archive and Library; The British Library, especially the Trade Literature archivists; Catalyst: Science Discovery Centre and Museum; Middlesbrough Institute of Modern Art, especially Elinor Morgan, Olivia Heron, Helen Welford, Connor Lagus and Holly Willats; Khadim Hussain; Leila Nassereldein and everyone on the CHASE-funded jaunt to Redcar and environs in 2020; Iris, Mordecai, Ben for company on the trip to Redcar, Middlesbrough and Marske and the thousands of people who keep the world of ICI alive in online threads, local and industrial history societies and the everyday banter on the streets.

Contents

Synthetic Beginnings — 1
Wonder Stuff: A Preamble — 2
Food and Death — 10
Company Origins — 18
Brave New World — 23
Nature and the Elements — 43
Futurology — 49

The Highpoints and the Low Ones: In, Over, Around and Under the Chemical Factory — 61
Who is Making Those Clouds? — 62
Old and New Nature — 72
A Plan — 79
Sociology and Chemistry — 85

ICI and the Senses — 93
End of Days for ICI — 94
Grasping at Threads — 113
News About Clouds — 118
Toxic Ground — 125

History as Synthesis — 137
Blueprint History — 138
Sources — 140
Method — 144

Chemical City: Origins 148
A Last Chorus 151

Index 157

List of Figures

Synthetic Beginnings

Fig. 1	Entries in ICI workers' photography competition, ICI Magazine, November 1935	5
Fig. 2	ICI's stand at Anglo-Palestine Exhibition, ICI Magazine, July 1933	12
Fig. 3	Soda Ash photograph, ICI Magazine, July 1933	14
Fig. 4	ICI Magazine, contents page, July 1933	17
Fig. 5	A photograph of 'Ammonia Avenue' from a July 1933 issue of ICI Magazine	24
Fig. 6	Photograph of sodium carbonate from ICI Magazine, July 1933	27
Fig. 7	A view of Billingham's hydrogenation plant from ICI Magazine, February 1938	28
Fig. 8	Alkathene advertisement from the 1950s	38
Fig. 9	Vynair advertisement, 1959	42

The Highpoints and the Low Ones: In, Over, Around and Under the Chemical Factory

Fig. 1	A new product, Monohydrate Crystals, showcased in ICI Magazine, July 1933	66
Fig. 2	The integrated company, from ICI Magazine, February 1938	69
Fig. 3	Alkathene advertisement in the 1950s	74
Fig. 4	Bottle of Trichloroethylene, England, 1940–1960 (Science Museum, London)	89

ICI and the Senses

Fig. 1	An artificial fibre ICI tie promoting Fertilisers	96
Fig. 2	The view from Wilton across the area	113
Fig. 3	Gulls on the beach at Marske by the Sea in February 2022	123
Fig. 4	Wind farm off the coast of Redcar in January 2020	126
Fig. 5	Sea off Redcar, February 2022	128

History as Synthesis

Fig. 1	Entertainment in the February 1938 edition of the ICI Magazine	139
Fig. 2	Safety advice in the ICI Magazine, February 1938	152

Synthetic Beginnings

Abstract This chapter establishes some of the history of the industrial production of synthetic materials. It considers the transition from the discovery and production of natural plastics to artificial or laboratory-made ones. The broader context of the industrial revolution in Great Britain narrows to consider the specific history of one company, Imperial Chemical Industries (ICI), and, in particular, its Tees valley factories, understood as dependent on specific features of landscape and weather in the region. The company's origins in war—its role in synthesising nitrogen in emulation of the Germans, for example—are explored. Issues related to the nutrition of growing populations are outlined. The impact of the Second World War on invention is drawn through the immediate postwar history of the company, which tracks the emergence of synthetic materials into everyday life, in the form of such things as artificial fibres for clothing, various forms of household plastic, household paints, brands which become household names. A press-button age is proclaimed in the company magazine. Links are made between these ephemeral, futuristic, easy commodities and the elemental matter and processes that underpin their production. A hint of the everlasting nature of their matter is brought up—the substances never go away, but at the same time, they deteriorate and leech into the environment.

Keywords Artificial fibres · Billingham · Fertilizer · Germany · Guano · Ironopolis · Nitrates · Plastics · Teesside · Weather modification

Wonder Stuff: A Preamble

Synthetic concoctions were the technical desiderata of the twentieth century. Through chemical reactions in factories, synthetic chemistry made molecules. It built new matter, imitating old matter, making synthetically forms of matter with qualities found in natural counterparts. Synthetic chemistry involves the use of chemicals and processes to force reactions, crack apart substances, induce binds and compounds and make products. Synthetics are replacement things, which are more or less like what they substitute. The structural codes of stuff—urea, acetic acid, indigo, glucose and more—had begun to be elucidated in the previous century and their synthesis worked on. What other molecules could be replicated, coaxed into being by means of synthesis? In pockets of the world, a quest began to produce, through synthetic means—that is, chemically, perhaps out of gases, coal and oil, water—substance and matter that was available in nature, but lacking for one reason or another in particular spaces and at particular times. Perhaps supplies were absent, if only temporarily, as a result of war or loss of territories. Perhaps someone realised that it might be cheaper to cook up useful substances in massive amounts in the laboratory. Perhaps, the argument ran, it was rational to be able to get hold of useful substances in controllable ways and at the wished-for quantities. Perhaps it was deemed right to use the advancing means of science and technology to make more stable forms of whatever was required. Perhaps the synthetic versions were more effective, generative or usable than their naturally occurring relations.

Plastics was one field of synthetic activity. When we talk of synthetics, we tend to mean the substitute substance or thing, rather than the process of synthesising. Synthesising ammonia or methanol, syngas, urea, acetylene, butadiene or aluminium sulphate may be hard to imagine for those not involved: synthesising happens through processes such as distillation, filtering, reduction, decomposition, hydrogenation and catalysis. But some of the resulting materials that use these synthesised products, themselves synthetic substances, are well known, and known sensuously. We know the taste of bananas and we recognise artificial banana taste. Many of us now know what artificial meat tastes like. We may be able to discern differing hints in artificial sweeteners. We ingest synthetic vitamins without question, and ride on synthetic 'rubber' tyres. Synthetic plastics are well known. They entered the world loudly, brightly and extensively, and have not left—some never will. We all have some sense of the difference between wool and its simulant cousin acrylic, made from petroleum or coal-based chemicals, or silk and its near-double, viscose, made from cellulose, or nylon, made from carbon-based chemicals, and unlike cotton, for which it was a surrogate. We can feel on our fingertips the differing qualities of ceramic, from clay, and melamine, made from urea.

The label, plastics, has been applied to any substance that was capable of being shaped or modelled, formed or deformed, upon exertion of an external force, be that heat or pressure or a chemical reaction. Nature makes its own plastic substances, such as pine resin, pitch, tar, beeswax, tallow, frankincense and myrrh. Each of these substances has been useful to humans. Each is also resonant with myth, legend and fantasy. Frankincense and myrrh were presents for the baby Jesus; Icarus flew near to the sun on beeswax and wings; the Pitch Lake of La Brea in Trinidad was legendary for punishing misdeeds; pine resin was burnt by Pagans at Solstice Festivals. Amber and copal, two fossil resins, stem from the back of time and, as they moulded themselves into manifold things, so stories have been moulded from their malleable substance. Amber were the tears of the Heliades, when they came across their dead brother, and Freya wept tears of amber too, as she searched for her lost love Odur; These are old and natural plastics.

As the industrial revolution whirred and crunched its way through the world, exploiting imperial possessions, seizing the resources to be extracted from them, organising agriculture and mining around utilisable and profitable substances and matter, natural thermo-plastics were discovered. The carapace of sea turtles, tortoiseshell, could be shaped and formed, once boiled and ironed flat, into jewellery boxes, furniture inlays or book covers. European Alpine Goat's horn was highly prized, for it could be pressed by iron moulds into buttons, combs, shoe buckles and handles, in imitations of ivory or pearl. Gutta-percha, the latex of the Southeast Asian tree, palaquium gutta, moulded knife handles for manufacturers in England and coated a submarine cable from Dover to Calais in the 1850s. Through this cable, a world-spanning network of communication was inaugurated. Vulcanised latex rubber was promoted at the end of the nineteenth century, harvested from vast plantations in British Malaya and the colonised areas of Africa, and used for tyres and boots. Scientists worked on these natural plastics, isolating their components. They deciphered their chemical codes and were soon able to make synthetic versions in the laboratory. There was a synthetic ivory, Parkesine, named after its inventor, Alexander Parkes, in 1862, made of cellulose and bent into form, once heated. From this came Xylonite and, then, celluloid, in 1869, developed to replace ivory billiard balls. Celluloid was a volatile substance, prone to explosion. Milky Galalith appeared in 1897. Bakelite—which is known as the first truly synthetic resin, the first synthetic polymeric material—was made in 1907. This 'material of a thousand uses', as its inventor, Leo Hendrix Baekeland, dubbed it, made its way into the lives of millions, in the form of telephones, whistles, saxophone mouth pieces, pistol grips and buttons.[1] Bakelite came close to people, mingling with their senses. From it came radio sounds that nuzzled the ear. It entered the mouth as a tobacco pipe. Into the telephone's shell-like receivers were sent whispers. This browny plastic was intimate with the consumer of the early twentieth century. When its patent expired, Catalin appeared, in 1927, with the added bonus of availability in a range of colours. After seeing products at the 1915 British Industries Fair, the English Queen Mary ordered and wore Casein jewellery, ennobling a material made of milk and formaldehyde.[2] The names of new materials clunked from lips like new hexes. Familiar and cooked up substances were whirled together to produce, rapidly, ever-new substances out of which old objects—buttons, combs, door knobs, necklaces, bangles—might be re-formed, and new ones—plugs, telephones, radios—given shape. What began as limited in colour—the white of ivory, the gingery brown and black of tortoiseshell, the orangey-gold of amber—fanned out in a rainbow of colours (Fig. 1).

Plastic provided the materials for photography, so that this synthetic world might look itself in the face. Cases for cameras and frames for photographs were made of the new synthetic supplies. At the end of the 1880s, the first transparent film roll was produced and a plausible use for celluloid was found. Celluloid adapted itself to other uses too. In the first years of the twentieth century, the Dreyfus brothers from Basel picked apart the plastic material of moving film stock—cellulose acetate—and adapted it for the British government as a varnish for protecting the structure of early aircraft. When war finished, it was re-moulded as artificial fibres. This was another area of intensive invention. Theirs resisted dyestuffs, but others followed that could absorb colours. And there were new surfaces such as Formica laminate, invented in Cincinnati in 1913, as an electrical insulator to replace the silicate mineral mica—'for mica'. By the early years of the twentieth century, plastics were in dining rooms and parlours, bathrooms and kitchens, in playrooms and gardens, in factories and offices, in tourist sites, on the battlefield and in the cinema.

Fig. 1 Entries in ICI workers' photography competition, ICI Magazine, November 1935

Once upon a time, plastics had been solely made by nature. Then, natural substances had been coaxed, through human invention, into forms. Finally, nature was outbid in the laboratory, in the creation of flexible and mouldable materials that could appear in any shape and any colour. The ability of humans to augment and replicate and outdo nature was extended, through technical rationality. This amplified a process of extraction of materials needed to produce and reproduce all that can be manufactured at scale in factories and their various plants. It commutes, in new ways, human labour power too into an extractable substrate. Through various, sometimes brutal, processes conducted on nature, including human nature—extremes of heat and pressure, for example—the world was changed from its social relations down to its molecules.

There were many sites in which synthetic production took place, but, in Britain, one area became a place of particular concentration—Teesside.[3] The names of some of the towns in the area—Billingham, Stockton-on-Tees, Middlesbrough, Wilton—became synonymous, across the twentieth century, with the company Imperial Chemical Industries.[4] ICI had other sites across Britain and across the world, but Teesside was the place of its birth and seemed to remain at its heart, until the heart stopped beating. And this company, ICI, would be a bellwether company—that is to say, a firm that indicated the wider health of the nation's manufacturing base.[5] Even in its demise, through its dwindling fortunes can be tracked the fortunes of some things that are much bigger than it alone—industry, economy, nature, labour, dreams and hopes.

The Tees valley provided a landscape that was crushed and melted and blasted and exploded, in order to bring about a new chemical world of synthetics. The region's plentiful deposits of coal and iron ore and minerals, such as anhydride, seemed to seal a fate. The landscape produced an economy, a social system, a mode of life and everything associated with that, even if that landscape became transformed into something other than itself. It was a place ground up and reconstructed by science and technology and then exported across the country and across the world, finding its way into people's homes, onto their bodies, into their bodies, as food, pharmaceuticals and so, while also resourcing more abstract imperceptible chemical transactions that are crucial to the processes of industry. The Billingham site on the north bank of the Tees river lent itself to these processes by its very features that have swirled and buckled for millennia: its water that gave good access and plentiful supply for industry, its coal for heat and transformation, its seam of the mineral anhydrite. But it is not just a landscape that remakes a world synthetically. It has to be coaxed out and fought over, processed and packaged.

Synthetics was not the only industry that developed in Teesside. Along the banks of the River Tees and close to the North Sea, various industries took root, including shipbuilding, coal and ironstone mining, steelmaking and, in the twentieth century, open cast strip mining and heavy engineering. Close to water, anything that was made could be transported easily across the nation and across the world. As industry and its products and its waste accumulated there, as people accumulated, drawn in to make and disperse and consume the outputs, so too did capital. Once synthetics came, Teesside was already a prime site of capital accumulation.[6] It was already a main centre of the capitalist world born in the nineteenth century.[7]

Synthetics came as a modern industry. The previous century's laissez-faire capitalism had left its mark and shaped the landscape. The railway had first hurtled noisily, steamily across the countryside, knitting it into industry and city, in the 1820s. The railway came to the marshlands and meadows, beginning with Locomotion No. 1, a steam train manned by its engineer George Stephenson for its first journey on 27 September 1825. The line, designed to transport minerals, ran from Stockton-on-Tees to Darlington and, later, in 1830, on to Port Darlington by Middlesbrough to provide coal staithes from the Eston Hills, Cleveland.

Sir George Head collected his thoughts on its early outings in *A Home Tour through the Manufacturing Districts of England, in the Summer of 1835*.

> The sound of the engines, on the Stockton and Darlington railroad, may be distinctly heard on a still day at the Dinsdale Hotel, like the flapping of mighty wings, as they pass along; and the line being in many parts circuitous, the puffs of smoke appear here and there among the trees in a thickly wooded country, enabling the spectator to mark the progress of the trains and trace their direction. In one part of the railroad the rails are laid straight for more than a mile together. Here I used to feel much gratification, by seating myself to watch the approach of the several heavy trains of coal-waggons, on their way backwards and forwards, laden and unladen, between the Darlington coal-field, and the staiths at Middleborough or Stockton.[8]

The railroad changed relations around it, he argued. The public were now no longer able to walk in its environs, but could speed through the same paths.

> On the banks of a canal navigators and loiterers infest the towing-paths and create a nuisance, but all descriptions of travellers on a railroad may rather be compared to a flock of pigeons or swallows, that confine their flight to the regions of the air, and leave neither track nor trace behind. Silence and stillness reign within its precincts, and harmonize with the grandeur of the spectacle; the rails converging in perspective form the track of a terrestrial zodiac, - lines terminating in points in the horizon, whence at prescribed periods earthly objects rise and perform their transit, while many a muscular arm toils in preparation for the phenomenon, which appears and passes away. As train after train of rolling wagons approached, a black speck first appeared in the distance, gradually and by slow degrees extending its dimensions; meanwhile the sound, like the roaring of the sea, became as a heavy gust of winds, and then, as the carriages receded, grew again less and less audible, till it expired in a low gentle murmur.[9]

The passengers in the train were like flocks of birds, moving as a murmuration, a zipping unit darting through the steam pumping from the train, which was a mobile factory for the manufacture of new clouds. The train was a machine, a factory on the move, making clouds so that more factories could transport matter into commodities. But it was reinvented nature too, a new type of bird, flapping its iron wings. It roared like the sea, thundered like strong winds, then passed away with a gentle murmur. This new nature ripped through the land, changing everything. Its drivers had given over their agency to the engine: 'Impelled by a power called by themselves into action, their arms folded on their bosoms, as if either lost in their own reflections, or dozing life away, they passively reclined in an easy posture'. The old cart horse, replaced by a new being 'endued with animal breath and progressive motion', was dragged behind the coal-waggons, no longer the power in front, but rather an entity subjected to speed, NEWLY conscious of the potential danger of the possible accident that had been brought into being.

> The sagacious animal, thus left to himself, on a bare platform of boards, within a couple of feet of the ground, and without side-rail or guard of any description, displayed a consciousness of the danger of jumping out,

by the mode in which he cautiously rested on his haunches, prepared by his attitude against the sudden possible contingency of a halt.[10]

The railway was the presence that catalysed industrial development. Stockton-on-Tees was a market town, built on wool exports, shipbuilding, rope and sail making. Its availability as the place of the first railway line was secured by the level nature of the land, with no steep drops and a stable geology. By extending to Middlesbrough, a place that thirty years before had just a handful of inhabitants, the railway and its affordances brought about a new town, poised in the middle between coal deposits and iron ores. Growth took off, especially after the discovery in 1850 of the main seam of Cleveland iron ore.[11] By the end of the century there would be 100,000 in the city.[12] On the damp marshlands, Ironopolis grew up, after a technical way had been found of removing the high level of phosphorus in local ore, which stymied its processing. Middlesbrough, a city now, had another name. It was dubbed, by Gladstone, the Prime Minister of the time, who visited in 1862, 'an infant Hercules', and this tiny hero would be one who, with every cry and whine and whistle, breathed out ever more carbon gases into the atmosphere. Beyond accumulation for some, what were the benefits of this expansion for those who made the materials that underscored and enabled the power of industrial shareholders? The workers who migrated there lived beneath a sky that was pierced by the tall chimneys required by the regional Cleveland Practice of iron production.

> The Cleveland blast furnace was developed specifically to smelt large quantities of relatively low grade ironstone as cheaply as possible, and to achieve this, reliance was placed on obtaining maximum thermal efficiency by increasing the height of the furnace stack in order to utlize the heat generated at the base of the furnace to heat the materials being charged in at the top.[13]

These chimneys rose to a height of eighty feet. Bumping and whizzing around beneath them, like free atoms, released from rural oversight and tight bonds, were the migrants drawn into to mine and bash the ironstone, forge pig and wrought iron and feed the stacks, twist and turn outputs into iron bars and rods, hoops and sheets and rails and plates and angles that were so important for projects, such as expanding the railroads in the US, Russia, Australia and India from the 1870s.[14] They came only for a short while, these young men who had gained their skills elsewhere, and worked in tough jobs to make some money and move on, giving way to new incomers. Accidents in the iron industries were common. Accidents bring new institutions into being. Hospitals, one paid for through worker deductions, in the form of insurance, one paternalistic and charitable, opened in the city. As steel replaced iron, and competition from Germany and the US emerged, a decline in Middlesbrough's economic fortunes set in the 1880s. Repeated visits by medical officers tried to establish the causes of high mortality rates, including that of infants in the area.[15] To be there was to be subjected to the brutality of industry, to its processes on nature, including on humans as nature.

Food and Death

The chemical industry of the twentieth century was concentrated on two products, textiles and fertiliser. These twin outputs clothed bodies and helped to feed those bodies, increasing numbers of bodies, urbanised, massified, industrialised and modernised bodies. Textiles and fertilisers angled chemicals towards life and living, warmth and food. It was the work of the twentieth-century chemical industry to enhance the availability of all this stuff by producing synthetic versions or substitutes. Textile factories had been the drivers of nineteenth-century capitalism and existed at the centre of a worldwide system of exploited labour, slavery, imperial possessions, including plantations and markets, mechanisation and struggles over the length and conditions of the working day.[16] Alongside the factories of cotton and wool developed, and with increasing speed, a parallel world of artificial fibres, such as Chardonnay silk and viscose rayon—processes inaugurated by a chemist dipping a needle into liquid mulberry bark pulp and gummy rubber, extruding it, silkworm-like, through a small hole, and depositing it in a coagulating bath. Viscose came in 1892, from wood pulp, perhaps from pine trees or bamboo, mixed with caustic soda and carbon disulphide, which produced a honey-like thick liquid with high viscosity. Artificial threads, with new names and laudable qualities, followed on. Proteins were treated chemically and known as regenerated protein fibres, one patented, for example, from gelatine, and experiments were undertaken on eggs and blood albumen.

The British Committee on Industry and Trade reported in its 1928 *Survey of Textile Industries: Cotton, Wool, Artificial Silk* on the advantages of this new 'smooth', hairless, scale-less product, benefits which included its softness and the lustre of the fabrics made from it, and furthermore: 'Owing to the smoothness of the filament, cloth woven from it will not readily catch dirt'.[17] Walter Greenwood's novel, *Love on the Dole*, from 1933, portrays bestockinged women workers leaving a textile factory in Salford, in the North of England:

> Clatter of clogs and shoes; chatter of many loud voices; bursts of laughter. Hundreds of girl operatives and women from the adjacent cotton mills marching home to dinner arm in arm, two, three, four and five abreast. They filled the narrow pavements and spread into the roadway.
>
> A generation ago all would have been wearing clogs, shawls, tight bodices, ample skirts and home-knitted, black wool stockings. A few still held to the picturesque clogs and shawls of yesterday, but the majority represented modernity: cheap artificial silk stockings, cheap short-skirted frocks, cheap coats, cheap shoes, crimped hair, powder and rouge; five and a half days weekly in a spinning mill or weaving shed, a threepenny dance of a Saturday night, a Sunday afternoon parade on the erstwhile aristocratic Eccles Old Road which incloses the public park, then work again, until they married when picture theatres became luxuries and Saturday dances, Sunday parades and cheap finery ceased altogether.[18]

Artificial fibres clung close to the body—those who still worked processing the old-style materials of cotton showcased on their legs the newer ones and, in their leisure time exhibited, temporarily, some carefree consumption of modern joys—each woman a bellwether of trends in the making.

Natural nitrates are made in soil by microorganisms called diazotrophs, bacteria such as azotobacter and archaea. They are made in the air too by the smash of air's nitrogen with the electricity of lightning—the nitrogen is fixed, converted into nitrogen compounds available for biochemical processes. Industrial modes of fixation were sought in the twentieth century. The task of the chemical industry was to emulate natural substances and processes, so as to generate more cheaply, more quickly, chemical compounds that could do the work necessary for expanding industries. Through synthesising processes and chemical manipulations, operations might be scaled up, reaction times optimised, procedures rationalised. This chemical work might be a work of speed up, catalysing quicker reactions, and it might condense space, making through synthetic trickery what could not be brought in from far away. Such was the work of ICI, which came into existence in 1926, following the merger of Nobel

Industries Ltd, Brunner, Mond and Company Ltd, the United Alkali Company and the British Dyestuffs Corporation Ltd. These companies combined to make a business of many parts, a coagulation, a concern, whose interrelated parts, whose spectrum of production, could beat all rivals and be a company not just for its titular Empire, but also for the world (see Fig. 2). In 1927, its first full year of business, it sold £27 million worth of products and made a pre-tax profit of £4.5 million. The new company needed symbols, such that it might be recognised across the world. Alongside a roundel with initials and wavy lines of the sea, adapted from one used by Nobel Industries Ltd, other company identifiers were employed. One was an upright strutting lion, bearing the stacked letters, I, C, I, in an approximation of some kind of scientific instrument—a winning entry in a 1928 company design competition, proposed by W. E. Anderson of the British Dyestuffs Corporation Ltd., Sales Department, Blackley, Manchester.

A speech by W.F. Lutyens, chairman of ICI Alkali Group, published in ICI's July 1933 magazine, gave a sense of early successes in certain

Fig. 2 ICI's stand at Anglo-Palestine Exhibition, ICI Magazine, July 1933

product lines. Great Britain, Lutyens observed, was, 'next to the United States, the greatest producer of alkali in the world, and of the total output of Great Britain Imperial Chemical Industries accounts for 99 percent'.[19] The alkali products to which he referred were soda ash, caustic soda, bicarbonate of soda and soda crystals and brand names 'Sesqui', 'Crex' and 'Pearl Dust' (see Fig. 3). The uses were multiple—for example, water was softened by alkali and it was used in shoe tanning and making chamois gloves. Lutyans' primary audience was businessmen, perhaps, those with polished shoes and fine hand wear. Alkalis made possible the development of artificial silk and viscose. For new fabrics, he explained, wood pulp was digested with caustic soda and carbon bisulphate. A thick yellow fluid was squirted through fine jets in an acid bath, making a thread that could be spun and woven into flexible material. Lutyens' talk set out a vision of how well-distributed alkali products were in the daily life of 1933.

> Before going to the office you go to the breakfast table. In the making of the table crockery, ash has been used and also in the glazing; your forks are electro-plated and ash was used in the process. If you are taking tea, bicarbonate may have been used to bring out the flavour of the leaves by adding it to the water, though this practice is not now so common as it used to be; the sugar if you use it, was refined with soda ash, or, if you use saccharine instead in order to keep down your weight, that is made from soda ash. Now we come to the milk: the bottle in which this came was scoured with 'Crex' or 'Pearl Dust'; your bread has been baked with bicarbonate of soda and previously the flour was matured with muriate and chlorine; the margarine - I do not allow you butter, you notice - has had quite a considerable amount of caustic soda used in its manufacture, especially in the refining of the oils from which it was made. Your new laid eggs have been preserved for several weeks in silicate of soda - caustic was used in the production of the aluminium of the pan in which your eggs were boiled. We then come to your marmalade: ash has been used by the manufacturers for scouring out the pans in which the marmalade (and jam) has been made. After you have gone to the office, your housework has to go on, and first of all your breakfast things have to be washed up, for which process either packet soda or soda crystals are used. Perhaps it is a Monday, in which case your clothes have to be washed; various alkalis are used in the washing of clothes, the strength or type of the alkali used depending on the material of which the clothes are made and their colours.[20]

The bath was galvanised with muriate. The torch was made with rice and caustic. The washing blue was made from ash and bicarbonate. The

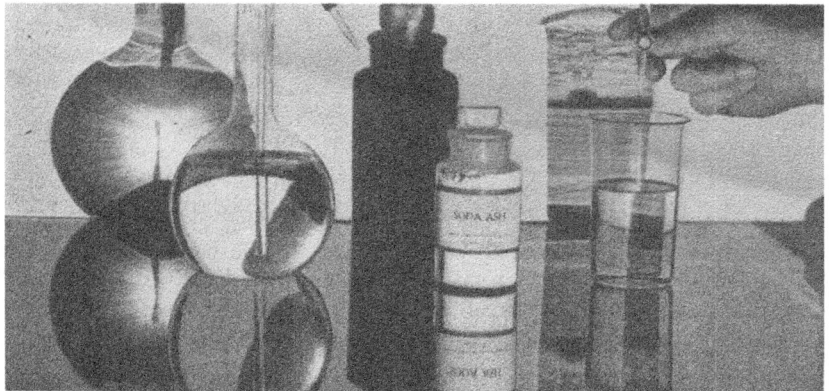

Fig. 3 Soda Ash photograph, ICI Magazine, July 1933

story of such predictable, conventional life, with its gendered division and vision of labour, its moves from production to social reproduction, continued. Pipes and tobacco and matches, the petrol and oil in the car, the road, the railway train, the office sandstone, the fire extinguisher, the coal in the fire, the artificial leather chair, the blotting paper, the ink: all required alkalis. The cardboard box, the telephone batteries, your table glass at lunch, the whisky, meat, potatoes, green, tin cans, raisin pudding, chocolate, and then the afternoon off, playing tennis on a hard court, solidified with calcium chloride, the racquet and ball were dependent on alkalis. If you took a fall, the plaster of Paris needed silicate of soda and the restorative snuff gained an additional bite from soda ash. And life itself and how it developed might be dependent on these chemicals. Lutyens closed his speech with an anecdote about a medical man in London who was experimenting with controlling the sex of children before birth. At a recent meeting of the International Conference on Genetics, it had been stated, reported Lutyens, that internal consumption of lactic acid would ensure a girl, while bicarbonate of soda should be taken, were a boy desired. He mentioned a German study in which, of seventy-eight babies born after ingestion of bicarbonate of soda, seventy-seven were male. 'In any case it is satisfactory to know that the Alkali group can now be regarded as being responsible for the future manhood of the Empire'.[21]

The speech appeared in the company magazine, where striking 'new objective' photography also appeared. This monthly magazine was a vehicle for producing an image and a sense of the company in its entirety, its parts bound together in Division reports, its workers kept up to speed on new developments, on safety measures and sales and coming events. Births, along with deaths, were part of the monthly reporting of the ICI magazine. In the same issue as Lutyens' address, as always, the tallies of newcomers—in proximity to figures on employee illnesses and deaths—were reported.

> Ammonia Group
> Congratulations
> To Bob and Mrs Gunn, a boy
> To Jack Jessop and Mrs Jessop, a boy
> To W. Matthews and wife, again, a boy.

Drikold Group and Sulphate Group reported on the birth of a slew of girls.

ICI was a culture in itself, an interconnected body of workers, plants, factories, Divisions, managers, families and social lives. The company was imperial. It strode a world, a whole world, parts of which were subjugated as colonies, but it was also enmeshed in its localities. The ICI company magazine reflects all this. In the 1930s and 1940s, it presented a concerned face. A body of workers had to be trained in a risky industry. Health and safety procedures had to be communicated wisely, and photography was used to offer lessons in how to operate around toxic chemicals, at heights or in proximity to fire. Accidents halted production and so halted the flow of commodities and so impeded profits. They were to be avoided. In return for the exercise of care in the factory, the company took care of life, running football teams and social clubs, outings and events, bringing into being management colonies and semi-detached houses for staff and foremen and rental homes for workers. Workers were invited to be productive in their leisure time too and to demonstrate, for example, how they had learned to look, with the best of their images printed in the company magazine or celebrated in photography competitions (see Fig. 1). In the notes section of the magazine, the workers were congratulated on their modest achievements in hobbies, their marriages, their births and were offered commiserations at their deaths.

The magazine issue of July 1933 brought together themes of workers' creativity, the risks of the plant and the social life of industry. A story titled 'Two Black Crows: A Billingham Tragedy' was the winning entry in a company story competition (see Fig. 4). Printed in the section titled 'Fact, Fiction, Leisure' and written by a worker, William Smith, at the Billingham Workshops Estimating Department, F. & Sp. Ltd., it told the story of two crows who liked to swoop down to pick up the garbage that lay around the plant. The workers on their occasional rests from 'the arduous toil of erection against time' would also listen to the sweet voice of fellow worker 'Jimmy', as he sang a song about the 'Two Black Crows'. On one rainy day, both crows landed on a wire and pecked at a second one, only to be electrocuted in a flash. The story detailed the events at a coroner's court assembled to decide the reason for the death. Evidence of suicide was lacking, and for murder there was insufficient motive. Accident could be ruled out as it was so 'unlikely in view of the precautions taken against such events happening at an ICI works'. The verdict had to be left open and it was hoped that 'by the deaths sufficient warning would be given to all and sundry to avoid live wires as they are apt to be dangerous'.[22] Or at least damage them only on fine days. A fellow worker who had to repair the crane in the rain contracted a cold—worsened by the shock received when his wife turned on the radio and he heard 'Jimmy' performing his song about the crows.

There was something specific about synthetics, something particular to the synthetic chemical industry. Close by, intermeshed within the chemical industries of ICI around Billingham, was the production of steel and heavy engineering, beneficiary of the iron found in the Cleveland Hills that overlook the Tees valley. A little further away in Tyneside, there was shipbuilding. These industries had something grand-scale about them—they made vast things, bridges that span the harbours and rivers of the world, ships that float across the seas, bigger than any building. But chemistry, the synthetic production that grew up in Teesside, made barely things. It made rather gases, chips of plastic, thin threads, powders, stuff from which to make dyes and paints, small things that fed into or fuelled bigger things. Things that were small might have been small enough to find their way everywhere, to distribute themselves like atoms or fleas and accumulate, forming environments, more or less perceived, fibres weaving into clothes, the fertiliser in the soil in which vegetables grow. At certain points in the twentieth century, Billingham, Wilton and North Tees held the biggest concentration of chemical plants in the world and there were

I.C.I. MAGAZINE

Vol 12 No 67   *July 1933*

CONTENTS

	PAGE
Frontispiece	2
Editorial	3
World Month by Month	4
In the News	5
Central Council Meeting	6
Anglo Palestine Exhibition	20
Among the Companies	22
Safety in the Works	24
Letters to the Editor	28

NEWS FROM OFFICE AND FACTORY:

News from Overseas	29
News from Headquarters	30

ALKALI GROUP:

Winnington	36
Wallerscote, Lostock	40
Middlewich	41
Silvertown	42
Fleetwood	43

BILLINGHAM GROUP:

Billingham	46
General Chemical Works	50
From the Departments	51

METAL GROUP:

Birmingham Hospital Saturday Fund	54
Kynoch	56
The Kynoch Press, Excelsior	57
King's Norton, Amal, Waltham Abbey, Lightning Fasteners	58
Hughes Stubbs, Selly Oak, Kingston	59
Landore	60
Sunbeamland	61

DYESTUFFS GROUP:

Hexagon House, Blackley	62
Huddersfield	64
Ellesmere Port Trafford Park, Grangemouth	67
Oliver Wilkins	69

NORTHERN EXPLOSIVES:

	PAGE
Ardeer	70
Westquarter	72
Regent	74
Roslin, St Rollox	76

CENTRAL EXPLOSIVES:

Gatebeck, Lowwood	77
Sedgwick, Gathurst Parr, Denaby	78

SOUTHERN EXPLOSIVES:

Faversham, Tonbridge, Tuckingmill	79
Lighting Trades Ltd. Earlsfield	80
Sutton	81

GENERAL CHEMICAL GROUP:

Central Laboratory Widnes	82
Castner-Kellner	83
Pilkington-Sullivan Gaskell-Marsh	84
Fleetwood Salt	85
Netham, Oldbury Stafford, Wednesbury	86

LEATHER CLOTH GROUP:

Hyde	87

LIMESTONE GROUP:

Buxton	88

PAINT GROUP:

Slough, Bordesley Green Stowmarket	90

FACT FICTION AND LEISURE:

Two Black Crows	91
Books	93
For I.C.I Girls	95
Films	97
Gramophone Records	98
July in the Garden	100
The Competition Page	102
Buy I.C.I.	103

Fig. 4   ICI Magazine, contents page, July 1933

so many things produced from the chemicals made there that the world of a certain period was saturated by its manufactured and exported matter. Cyanide from Billingham mined gold in the Arctic, Australia and South Africa. Products were sent from Teesside to Argentina, India, China, to everywhere. Plants were built across the world under the name ICI. Products were received from the whole world too. And all this was part of a competition, where the opponent was changing. But in the early days, it was Germany. German competitors had to be smashed and their markets taken. But there were always more competitors arriving on the scene—Italians with their alkalis, the Japanese with non-ferrous metals. Nothing could rest. Reactions and catalysts were constant. What counted as the centre of the world? Where is the hub of the imperial industry? It might be Billingham—at least as it figured in the striking diagrams ICI produced for trade fairs and in its company publicity. At least for a while. Such rankings had to be fought for.

Company Origins

What led to the construction of this chemical combination? An ICI report, dated 28 July 1939, and authored by key personnel at the company, recounted the early history. Titled 'A Short History of Billingham up to the Time of the Formation of ICI', it began with recognition of a contradiction: 'it is paradoxical that nitrogen, the element which is the basis of all life under the guise of protein, is also the basis of explosives'.[23] Wars, it argued, became more violent over time and the need became pressing for more means to produce more death, as supplies for gunpowder for 'the whole civilised world', transported from India, China and Egypt were inadequate for growing requirements or were blocked in their movements around the globe. In France, noted the report, during the Napoleonic Wars, nitrogen farms had been established, mining nitrates from the excrement of domestic animals. German chemist Justus von Liebig discovered, in the 1840s, how nitrogen acted as a fertiliser and so he underlined the life-promoting aspects of the element. The discovery of Peruvian guano, deposited by birds, to a thickness of 150 feet in places, fed industries of death, life and dyestuffs for a while. The English took commercial control of these resources. Chile saltpetre, or sodium nitrate deposits, were found in the Atacama Desert and led to war between Peru and Chile from 1879 to 1883, over their possession. Chile won and gained a monopoly on supply, but it was a dwindling resource. Needed for both death and life, the ICI report stated: 'Consumption has increased steadily since then, and the nitrogen industry, originating as an aid to war, is now essential in time of peace'. It cited Sir William Crookes' warning, in 1898, that nitrate deposits in Chile would deplete eventually and the rising population would not be able to be fed, because agriculture, without adequate fertiliser, could not produce enough food. Left to itself, or rather to what German scientist Justus von Liebig designated the 'robbery system' of agriculture, which in the name of economic expediency did not replenish nutrients in soil, there would be famine.[24] The world, stated Crookes, could never consist of more than two billion people and the possibility for endless expansion, new markets, new lives would be lost. The task from then on was to fix atmospheric nitrogen—the vast stores in the air were to be brought to earth and utilised.

German scientists found a way of synthesising ammonia—from nitrogen fixed from the air—at the start of the twentieth century. Ammonia could be converted into nitrites and nitrates, by bacteria in the soil, and thus fertilise abundance. Within a few years of setting its sights on the problem, Germany had become self-sufficient in nitrogen production, and an exporter of fertilisers. It appeared to have an endlessly augmentable capacity to use its manufactured supplies for war explosives. Britain wanted, or needed quantities of explosive material too, and had to work out how to deploy the same process, how to produce ammonia by a high-pressure synthesis (at 250 atmospheres) using something like the Germans' Haber–Bosch process.[25]

British research scientists attempted to emulate their work in the Ramsay Laboratories at University College London, and after some successes, were tasked with opening a commercial plant for 'the production of synthetic ammonia, nitric acid, and the required explosive ingredient ammonium nitrate'. A farm, Grange Farm, or Chilton's Farm, in the village of Billingham-on-Tees, was selected for the chemical works supplying wartime needs: 'the production on a large scale of nitrogen and hydrogen and for combining nitrogen and hydrogen so obtained for the production of ammonium nitrate to the ultimate extent of 60,000 tons per annum'.[26]

The Billingham site, expanded from an initial size of 250 acres to 800 acres, was agricultural land. It was chosen because of the availability of electricity from a nearby power station that began construction on the banks of the River Tees in 1917, the proximity to coalfields, the abundance of cooling water and the potential availability of skilled labour. Approved by the Ministry of Munitions on 22 March 1918, it was named the Government Nitrogen Factory—but factory construction was delayed, perhaps as a result of the German offensive of March 1918, when the production of poison gas was prioritised, or perhaps because of labour shortages and a lack of building materials.

British inspectors, benefiting from Germany's defeat in the First World War, toured German factories to pick up details of the processes. An anecdote records how, the day after the Armistice, Francis Arthur Freeth, chemist and director of research at Brunner Mond, went to London 'with an idea burning inside him'. The following conversation is said to have ensued between Freeth and John Fletcher Moulton, Director-General of the Explosives Committee and member of the House of Lords:

> Moulton (smiling): 'Well Freeth, what is it today?'
> Freeth: 'Sir, in three weeks' time the British Army will be on the Rhine, let us send a chemical commission and pinch everything they've got.'
> Moulton (in a stern voice but with twinkling eyes): "My dear Freeth, you must not propose a burglary to a Lord of Appeal. Go and see Sir Keith Price and let me hear no more of this'.[27]

And so, he did visit Price, Moulton's deputy and the head of chemical warfare production at Porton Down. The mission went ahead. Methods were conveyed back to Britain—though it is told that some crucial details were stolen in transit and had to be reconstructed.[28] It was decided to sell the work in progress, as long as the Government interests in the production of nitric acid for service explosives were safeguarded. The factory remained unbuilt and unequipped, but roads had been made and foundations laid and plant facilities ordered. The scheme was advertised in the press—though with no indication that 'information procured by the Government Commission to the German factory at Oppau would be included'.[29] However, applicants were told that more details would be provided upon application to the Ministry of Munitions: 'There were no suitable applicants other than Messrs. Brunner, Mond'.[30] The contract was signed on 22 April 1920, the company named Synthetic Ammonia and Nitrates Ltd was established on 3 June 1920, and the government provided the firm with all the details on nitrogen production that it had gleaned in its victory tours of Germany. Brunner Mond Ltd was to construct a 'peacetime fertiliser factory', under the stewardship of private industry, rather than government. By 1921, a successful process for the synthesis of ammonia from water gas and nitrogen had been worked out. Research engineers continued to work on the problems. Frank Ewart Smith was one employed as a research engineer in Billingham, from 1923, and he worked on engineering problems to bring about the production of ammonia, such as developing a gland, or pressure seal, for use on the piston rods of machines that circulated hydrogen and nitrogen, making it possible to increase the pressure from 200 to 300 atmospheres.[31] Eventually the staff were able to copy, or synthesise, the German ammonia plant at Oppau, Ludwigshafen. The Billingham production plant began operation in December 1923. Evidence of the success of the process was sensuous, tangible:

> The reader may gauge for himself the thrill of those operating the plant, when they first smelt ammonia in the synthesis system on that memorable Christmas of 1923.[32]

A 'History of Process Department (Nitrogen Division)' special report, from 15 April 1937, reinforced the sensuous proof of triumph in producing synthetic ammonia.

> Ammonia was first made on December 24th; after smelling it Col. Pollitt, Slade, Cowap and others went to Elton Hall to celebrate.[33]

A pilot plant producing ammonia was commissioned, but there was soon a fatal accident and the mechanisms needed more research and redesign—something Ewart Smith took on for six months. Systems improved. Ewart Smith included photographs from the Oppau factory, which served as the model, in his 1926 report on automatic weighing and bagging of sulphate, a project that began once enough chemicals could be produced daily. The photographs depicted small quadrangles with white mountains and metallic structures and they provided black and white evidence of the seemingly more efficient, that is to say, labour saving, processes in operation in Germany.[34] The intention was to move towards completely mechanical handling of bulk products, from manufacture to storage to bagging.

A vast supply of the mineral anhydride was discovered underneath the plant and used from 1928 for the sulphate plant. More discoveries meant more products, brand names and processes, such as the chemical fertiliser

Nitro-Chalk, in 1927, or the synthesis of methyl alcohol from hydrogen and carbon monoxide, and there was work towards making oil out of coal. In addition, engineered instruments capable of doing this work were key, steels for pipes and vessels that could stand up to the rises and falls in pressure and temperatures involved in the ammonia synthesis converters.

A plant had been brought into existence that could ensure Britain had a supply of explosives from nitric acid, should they be required, and ammonium sulphate, as plant fertiliser, would help feed the growing masses. This opened an important chapter in synthesising materials. And if it came too late for the Great War, there was always another war to be serviced. Smell was the proof that ammonia had been synthesised. Smell was evidence of impending doom too, in the First World War, when the smells of poison gases gave soldiers warning of what was to come: phosphene smelt like new-mown hay; dichloroethylene sulphide smelt of horseradish, and so was nicknamed mustard gas, but was disguised often disguised by a compound, xylyl bromide, which had a perfume of lilac. Fritz Haber helped develop these gases, gases which remade the atmosphere, weaponising the air in order to compete for the land that existed beneath the poisonous air. Air could give up nitrogen and that quarry be used for life or death. These gases, though, had more singular aims. It seemed unlikely that the deathly dynamic would decline. It seemed clear that the world existed only ever between the wars, not after them. The chemical industry combined in order to cover, as one behemoth, everything from birth to living to death, from enhanced production to vast destruction, from curing to killing.

A conglomerate of chemical companies, IG Farben, formed in 1925 from the merger of six chemical companies—BASF, Bayer, Hoechst, Agfa, Chemische Fabrik Griesheim-Elektron and Chemische Fabrik vorm. Weiler Ter Meer. This was a peacetime rival to British chemical production and in its image four companies of the British chemical industry had combined in 1926. In 1925, Walter Benjamin—or more likely his wife, Dora, who had studied chemistry—published a newspaper commentary, in the German *Vossische Zeitung*, on the coming war and the weapons of the future, which would suffuse the cities, new killing grounds, as gases, carrying the odour of violets. This perfumed deathliness would come as the result of a surprise aerial attack, against which there was no defence. The commentary was titled 'The Weapons of Tomorrow: Chlorazetophenol, Diphenylamchorlizine and Dichloroethyl-sulphide'. The 'coming war' would be fought chemically, and the article named the 'tongue twisting chemical vocabularies' of the gaseous killing tools manufactured by IG Farben.[35] In 1929, Benjamin named the conglomerate again, as he argued that the single certain thing to be sure of—the fate of the world—was that, in peacetime, chemistry prepared for war.

> Pessimism all along the line. Absolutely. Mistrust in the fate of literature, mistrust in the fate of freedom, mistrust in the fate of European humanity, but three times mistrust in all reconciliation: between classes, between nations, between individuals. And unlimited trust only in IG Farben and the peaceful perfecting of the air force. But what now? What next?[36]

Pessimism—all we know is that something is set in train and it will weaponise tomorrow. It will use chemistry as a tool of the state and business. That means it will capture the cosmic power of generation in order to remake a world.

Brave New World

A report in *Electrical Review*, dated 12/19 December 1930, introduced readers to 'a vast organisation served by one of the most remarkable electrical installations in the world which has never previously been described'.[37] 'The factory of Synthetic Ammonia & Nitrates Ltd, is certainly the largest chemical works within the Empire and must be one of the largest of its kind in the world', it stated.[38] The article mentioned how a dozen years ago 'the site had at its focus a small farmhouse, Billingham Grange, which still stands with neighbouring tree'. In the last few years, a systematically planned site had zoned production into areas, with nitrogen products in the north, alkali products in the south, cement works in the north east corner, near the river bank, where products could be loaded onto ships or rail. What was produced there came to name its streets, its byways. A huge apparatus loomed on Ammonia Avenue (see Fig. 5).

Production fertilisers and explosives were supplemented from the 1930s by pharmaceuticals, a development out of the dyestuffs industry. Attention turned as well to plastics. Bakelite had proven popular amongst consumers and it was clear that petrochemicals would soon yield plastics. ICI established a central plastic materials committee in 1927, and took over Croydon Mouldrite, where phenol–formaldehyde powders were manufactured. Plastics Division co-ordinated polymers research at various sites: Urea-formaldehyde at Billingham; vinyl resins at General

Fig. 5 A photograph of 'Ammonia Avenue' from a July 1933 issue of ICI Magazine

Chemicals; nitrocellulose in Ardeer, where explosives were made; and phenol–formaldehyde at the Dyestuffs works in Manchester. Finding new materials, substitute materials, was a priority. ICI filed a patent for polyethylene in 1936, and once preparations for war were underway, manufacture in huge quantities meant it could be used for the insulation of Radar equipment on warships and aircraft.

Around the factory grew a village, or garden city.[39] A village, a garden city, sounds attractive, as if nature were tidied up and made serviceable to residents. By 1936, 20,000 people lived there. Perhaps it was neater, more planned, more modern than the industrial settlements that horrified George Orwell at this time. These he described in *The Road to Wigan Pier*, published in 1937, result of his journeying through England the previous year, as 'a civilized man venturing among savages'.[40] He wrote, with some disgust, of the 'lunar landscape of slagheaps' in Wigan, where the factories belched out plumes of smoke and the canal path was mashed together out of cinders and frozen mud. All around were pools of stagnant water in ancient pits that had subsided: 'It seemed a world from which vegetation had been banished; nothing existed except smoke, shale, ice, mud, ashes, and foul water'.[41]

Worse still to his eyes was the sight and site around the factory works in Sheffield:

> Sheffield, I suppose, could justly claim to be called the ugliest town in the Old World: its inhabitants, who want it to be pre-eminent in everything, very likely do make that claim for it. It has a population of half a million and it contains fewer decent buildings than the average East Anglian village of five hundred. And the stench! If at rare moments you stop smelling sulphur it is because you have begun smelling gas. Even the shallow river that runs through the town is usually bright yellow with some chemical or other. Once I halted in the street and counted the factory chimneys I could see; there were thirty-three of them, but there would have been far more if the air had not been obscured by smoke. One scene especially lingers in my mind. A frightful patch of waste ground (somehow, up there, a patch of waste ground attains a squalor that would be impossible even in London) trampled bare of grass and littered with newspapers and old saucepans. To the right an isolated row of gaunt four-roomed houses, dark red, blackened by smoke. To the left an interminable vista of factory chimneys, chimney beyond chimney, fading away into a dim blackish haze. Behind me a railway embankment made of the slag from furnaces. In front, across the patch of

waste ground, a cubical building of red and yellow brick, with the sign 'Thomas Grocock, Haulage Contractor'.[42]

There is sensory delight, a kind of sublime terror, a 'sinister magnificence' to be had amongst this, where the drifts of smoke are rosy with sulphur and flames lick out of the chimneys. The boys are lit up in red as they haul 'fiery serpents of iron to and fro'. But mostly this industrial scene was an affront to the senses. Chemicals pooled from the factories. Soot and smut settled on everything. The air turned black. The company insignia made clear for whose benefit such a shambles existed.

Perhaps Billingham and the settlements surrounding ICI's factories promised more or something different, a new type of industry, well organised, contained. This region too had had its proliferating chimneys and furnaces and pipes and ovens, and it had chemicals that were seeping into the water, amidst the noise of production and the smells in the air. It made soda that cleansed the world, as much as it dirtied it (see Fig. 6). It would make more and many other things through synthetic means. Was this industrial future, with its array of new brand names articulating into the world complex chains of chemical reaction, something that was worth fighting for?

War's thunder rumblings could be heard before the lightning struck. By the mid-1930s, a government committee was exploring what would happen were there to be a large-scale military conflict in Europe. ICI, together with arms manufacturers, began planning for future conflagrations. Billingham's ammonium production facilities, crucial for the ammonium nitrate necessary for vast quantities of explosives, were a vulnerable target and so new production facilities were planned and the efficiency of the compressors for production improved (see Fig. 7). Fighter aircraft required high octane fuel and, in anticipation of increased wartime needs, in February 1939, Billingham chemists and engineers designed and constructed a plant, Heysham Aviation Fuel Works, in Lancashire. It was operated by Trimpell Ltd, a name derived from the three companies that had a stake in it: Trinidad Leaseholders Limited, ICI and Royal Dutch-Shell.

Through the years of war, ICI worked on developing new weaponry, including a defensive weapon that could knock out tanks, in the case of an invasion. Ewart Smith brought into production a shoulder-held, recoilless anti-tank weapon that was able to penetrate 4-inch armour-plate. It was rapidly designed, made and put out into active service in October 1942, in

Fig. 6 Photograph of sodium carbonate from ICI Magazine, July 1933

North Africa, under the name PIAT (projector, infantry, anti-tank). Ewart Smith brought together engineering design staff from Billingham, cajoled through submission to a staff assessment scheme that distributed promotions by merit. He also encouraged application and competition amongst the staff to design guns based on the most modern technologies, with faster rates of fire and greater abilities to pierce through tank metal. These engineers tried as well to divine the secret projects being carried out by the enemy. They examined photographs and documents from Air Intelligence to work out if the enemy was designing long-range rockets and flying bombs. George Whitby, an ICI man, studied photographs regularly and perceived, in April 1944, the shadow of a rocket at Peenemunde, in East Germany. It was apparent to him that some new mechanism of firing was in play. Fragments of one of the German test rockets that landed in Sweden, in June 1944, were analysed by Ewart Smith and Whitby, who knew about fuels, centrifugal pumps, mechanical and electrical control

Fig. 7 A view of Billingham's hydrogenation plant from ICI Magazine, February 1938

gear and supersonic flight. The two men were able to dream an equivalent rocket into being, as an exercise in reconstructive engineering and visual analysis.

Inside the factories in wartime Britain there was secrecy too, secret even to those who worked in proximity, as the future was worked on and foreshadowed in experiments and clandestine inventions. This future that had been in preparation came to light only later. In the ICI magazine from October 1957, a report was titled 'Atomic Energy Disclosures'.

> One night during the war a worker in a forging shop at the Witton factory of Metals Division bet his mates they could not lift the billet of metal he had been working on. Several of them accepted the bet, for the billet was only 7 in. in diameter and 8 in. long. To their consternation, none of them could raise it even an inch from the floor.
>
> This was not surprising, for the billet was of uranium - a metal of a density far beyond their experience and weighed 200 lb.
>
> Few laymen in those days had even heard of uranium, but at Witton billets of it produced at General Chemicals' Widnes plant were being experimentally rolled, extruded and forged under conditions of secrecy as part of ICI's contribution to the atomic energy project.
>
> The full story of this contribution has now been told for the first time. Writing in The New Scientist, Mr. D. R. Willson of the Atomic Energy Research Establishment gives details of the work done by universities and industry up to 1946.
>
> The Divisions of I.C.I. chiefly involved were General Chemicals, Metals and Billingham. At Widnes Europe's first metallic uranium was made in 1942. Later most of the uranium needed for the Harwell piles was made there.[43]

If atomic work remained obscure, made explicit after the war were the apparent benefits of other aspects of wartime invention. An ICI exhibition, from 5 to 28 June 1946, at The Tea Centre on Lower Regent Street in London, was titled 'Chemistry at Your Service'. It showcased those chemical achievements of British research in the war period, which, it was claimed, would improve life in peacetime. The foreword to the brochure observed:

> War is a great stimulus to many kinds of scientific research. During the last war the British chemist gave convincing proofs that his resource and inventiveness are second to none.

> This exhibition presents eight of his outstanding discoveries, either made or developed in this period. To have attempted to cover the whole wide field of British chemical research in one exhibition was impracticable. The method has therefore been chosen of focusing attention on those examples which are both notable in themselves as major scientific achievements and hold out promise of benefit to mankind.
> It is of interest to note how in chemistry, as in other spheres, the developments of wartime have even greater uses in time of peace.[44]

War's challenges, the inventions designed to spread or mitigate horror, find some sort of justification in the days of peace, or in the days between the last and the next war. ICI put on display a number of chemical substances—their efficacy was to be used both domestically and abroad, in still-existing colonial possessions and dominions. There were sophisticated insecticides, including the 'most deadly insecticide ever discovered', Gammexane, which was assumed to be 'practically harmless to man and animals'. It was 'used against all manner of insects, from bed-bugs, fleas, lice, and cockroaches to the mosquito (carrier of malaria, yellow fever, and dengue), the tsetse fly (sleeping sickness), and the African migratory locust'.[45] In smoke pellet form, once ignited and exploded like a bomb, it could disperse itself, as a cloud and settle on walls and surfaces, rendering them lethal.

Drugs and insecticides and fertilisers were signs of progress in the battle against problems of pests and disease, especially in colonies, where malaria was rife or insects threatened cash crops. Substitute substances, though, remained substitutes, not quite as good as what they replaced through laboratory invention. Plastics were showcased in the exhibition, artificial forms of lowly status—they were described as complementary to traditional natural materials.

> The meteoric rise of the plastics industry is mainly due to the research chemist, with his constant striving for new or improved materials. Today, plastics loom large in the popular imagination. There is indeed a danger that their potentialities may be exaggerated. The truth is that the established materials - glass, wood, steel, and concrete - will always be first choice for the purposes for which they are best fitted. Plastics are complementary.[46]

The 'real future' of plastics lay there where natural materials failed or in the conception of entirely new applications. Two new plastics 'were brought to perfection by the exigencies of war': 'Perspex' and polythene.

> Methyl methacrylate, the material from which 'Perspex' has been developed, was originally produced in 1930 as an interlaying medium for safety glass. For this purpose it proved a failure.
> Methyl methacrylate has, however, the peculiar property of 'polymerising' - that is to say, its molecules tend to link up in long chains to form 'giant molecules'. This polymerised form is 'Perspex', which possesses outstanding transparency and lightness, and will neither splinter nor shatter like glass. Naturally, during the war it found its greatest use as the glazing material for the noses, shields, and turrets of fighting aircraft.
> Important as these uses were, 'Perspex' has infinitely greater potentialities for domestic and industrial purposes. Today it is being used in corrugated sheets for roof-lighting, in prisms and lenses, for surgical and dental equipment, and for an apparently boundless variety of fancy goods for the home. While surpassing transparency is one of its greatest attractions, it is beautiful in a range of translucent colours.
> Industrial artists are beginning to realise the unique charm of 'Perspex', and to use it as the medium for entirely new designs.[47]

Polythene was presented in the 'Chemistry at your Service' exhibition as 'one of the greatest discoveries in plastics of the past decade'. It too was a polymer, of ethylene, a constituent of coal–gas. Under extreme pressure and high temperature, it was forced to polymerise in 1933, in an explosive, difficult process.[48]

> The development of polythene came at the very moment when such a material was urgently needed for Radar. The first plant began production on 1st September, 1939 - the day the Germans invaded Poland. Before it had produced a single pound, the order was given to double its scheduled output. The great technical superiority of British Radar over that of enemy nations was largely due to polythene. Without it, Radar could never have been brought so quickly to perfection.
> Polythene also found wartime uses in telephones, radio and submarine cables, and as a waterproof packing for sensitive drugs, such as mepacrine.
> In the ampler days of peace, its uses will be correspondingly greater. It is already being used for lamp-shades, handbags, acid-proof containers, funnels, and piping, and other new applications are being reported almost daily.[49]

Other inventions of British chemical research included clothing made of artificial fibres, which took advantage of the abundant produce of the British Empire. 'Ardil' was one, made from peanut waste gleaned from the 8 million tonnes per annum of nuts imported to Britain from colonial territories. Originally this waste was used as animal feed but ICI scientists found that the protein could be dissolved in solutions of urea and sodium hydroxide to denature and unfold the protein arachin. The result could be wet-spun to form fibres that were wool-like.[50]

> The sheep produces its wool by long and complex processes from the vegetable proteins on which it feeds. So far as wool production alone is concerned, this is slow and wasteful, since only a small percentage of the proteins consumed by the animal is converted into wool. The chemist has now succeeded in by-passing the animal and producing wool-like fibres direct from the monkey nut or ground nut that grows abundantly in many parts of the British Empire.
>
> The protein, consisting of long chains of amino acids linked together, is extracted from the nut and made into a solution. This is forced through the fine holes of a spinnerette into a bath, where it thickens into fine thread. It has been named 'Ardil', because it has been developed at the I.C.I. laboratories at Ardeer, in Ayrshire.[51]

'Ardil' was no equivalent substitute for wool, but rather seemed second-best. Science created a parallel world, not quite ready to step in and take over, holding back, not yet good enough or valued enough to remake what had gone before it.

The conclusion of the exhibition brochure observed how chemistry was an ever-progressing field of activity. There were always 'greater developments tomorrow'. The ICI writer evoked the image of Penicillin, 'evolved from a mould': 'There are many kindred moulds, each representing virgin fields of research which await the investigator'.[52] What moulds, what mushrooming ideas might be made tangible and change the course of history? Chemical research had cracked time and overcome space in the laboratory, but it also brought its own accelerations:

> All the discoveries included in the exhibition were made by British chemists within a period of no more than fifteen years - 1930 to 1945. No other nation can point to such a record of chemical achievement. Today British chemical research stands on the threshold of a new age, holding who knows what discoveries for the future benefit of mankind.[53]

Other materials moved from war-designated uses to home use. There was Holoplast, for example, made from raw materials from Billingham as an inflammable panel for partitioning in war ships. Made of Kraft paper doused in resin and subjected to heat and pressure, it found uses in the newly built schools and other buildings as internal walls and workbenches. It even became external walls and so marked out the separating spaces of the modern world. This panelling of laminated plastic, once furnished with a scratch-resistant, polished mahogany-style finish, after a thin band of aluminium had been heat-shrunk around its perimeter, to cover the honeycomb edge of the surface, was used in the late 1940s by furniture designer Ernest Race to make tables. These defined, with their tapering legs and pastel colours, the lighter design of the post-war look.

The war was over, but it might have seemed that its clouds never went away. The cover of the March 1950 edition of the ICI magazine showed a photograph of a broody sky of dark clouds, heavy with rain, black and dark grey, against a pale grey sky. Its title was 'Impending Storm'. Inside, news from Billingham included Dr. G.I. Higson, joint managing director of Billingham Division, arriving in Pakistan for an industrial mission.

> The mission's terms of reference are to explore and report to the Government on the steps which might be taken to assist the flow of trade in both directions between Britain and Pakistan, and particularly on the way in which British interests could assist further in the planning and execution of schemes which the Pakistan Government have under consideration for the economic development of their country and the expansion of trade.[54]

The processes of decolonisation and independence were forcing new economic settlements. Perspex found a curious use in economic modelling, in the field of hydraulic macroeconomics. In 1949, William Phillips built, out of Perspex and a hydraulic pump that he had salvaged from a Lancaster bomber, the mechanism for a modelling of economy by 'fluidic logic'. He called it MONIAC (Monetary National Income Analogue Computer), or the Phillips Hydraulic Computer or the Financephalograph.[55] Plastic pipes and tanks designated as sectors of the economy were attached to a two-metre-high wooden board. Differently coloured waters cascaded or pumped from one part of the mechanism to another through sluices and into troughs: flowing out to health and education, savings or investment, pumped back up to the treasury as taxation and all at varying speeds. In this way, a kind of Keynesianism in

Perspex visualised the economy, understood as a circulating and managed system. The economy might be like the civically maintained pipes of a city connected to bathtubs and reservoirs. What gushes through these pipes? What new materials might speed up economic circulation? What did the company that made the Perspex want of all this, as post-war discussions considered the future form of the economy as public, private, nationalised, shareholder, state-subsidised or in any other form?

If there was newness, there were also lingering fragments of the past. The shattering effects of war were still visible in bombsites, semi-collapsed houses, prefabs and allotment gardens, where small state food rations were supplemented. Abandoned military sites crumbled in the countryside, while others found new uses as nuclear shelters and command centres in an age of the Cold War, one in which Britain was participating by developing nuclear technology. Wars still happened—in which British conscripts fought, in Korea or elsewhere. And this Britain existed under a fog of pollution, the price paid for being the most industrialised and most urbanised country in the world—a quarter of world trade in manufacture in 1950. There were ships, cars, textiles and coal, all produced for export. Coal heating at home and in industry—refining oil and chemicals—smogged the air, attacked the heart and lungs. One city, Leicester, grew fat on nylon—for it was the centre of the hosiery industry.

Small details provide evidence of this growth that was celebrated. In the Billingham news section of the cloud-covered magazine issue of March 1950, someone was honoured:

> Mr. Norman Sturdy, senior chargehand of the Anhydrous Ammonia Filling Station of Billingham Commercial Works, has been awarded the British Empire Medal in the New Year's Honours List. He started work at Billingham nearly twenty-three years ago, and joined the Anhydrous Ammonia Filling Station in J 929. Last year the station's output reached the record figure of 4800 tons, of which 1830 tons went for export.[56]

Was it his record success, or something beyond and outside of him? Was it a success that was not his alone, but the company's, the state's and the country's?

Another wartime product from ICI was polyethylene terephthalate. In 1941, chemists John Rex Whinfield and James Tennant Dickson made a fibre that would become known as Terylene. Billingham provided the raw materials of paraxylene, ethylene glycol and nitric acid to make a substance that formed hard ivory-like threads, which were melted, extruded and drawn into fibres by the Plastics Division. So successful would this plastic fibre be, and was determined to be, that a huge new factory was built near Billingham, at Wilton, and it opened in 1949.

The area, its families, those drawn in from across the region and the country and beyond, thrived from production in the post-war period, when state-subsidised investment helped modernise manufacturing. Roads, schools, health facilities and homes for workers were built, especially if these met with the needs of ICI. The 1949 Outline Plan for the North East Development Area prepared for the Ministry of Town and Country Planning mentioned the expansion of ICI to its site at Wilton:

> The development of such a large unit as Wilton on Teesside, the one area where labour is in short supply, must entail some influx of population if its labour requirements are to be fully met.[57]

A working populace needed to be generated—for heavy chemicals was the key industry here and that was what must be served. It was not so much 'chemistry at your service' as 'you at the service of chemistry'. This required homes for workers, but ICI was not building company houses. Instead a 'gentlemen's agreement' was brokered between the company and Eston Council. A letter from Mr Gofton of ICI to Mr Basil R.W. Potter, Clerk and Solicitor to Eston Urban District Council, in 1947, stated:

> We appreciate the fact that your Council are not permitted to enter into a binding agreement to let these homes to our employees, and that the arrangements in this connection must therefore rest on the basis of a 'gentlemen's agreement' between the Council and the Company. If you will let me know when the first of the houses are nearing completion I will arrange to supply you with a list of our employees in the order of priority in which we would like the tenancies allocating.[58]

Those who had worked in now closed ironstone mines might be lured to Wilton and they would live there in council housing, courtesy of the agreements between gentlemen. The conjoined future of corporation and workers promised to be robust. A poster issued by the Teesside Industrial Development Board, in 1949, depicted Teesside on a map of Britain, its regional area was shaded in an orangey pink, and around it emanated concentric rings. It was as if it was a heart whose beating kept the rest of the country alive. The rippling rings stretched out over the sea. It appeared like rings on the surface of water through which a pebble had plopped. The sea was there, so much sea—sea to feed a sea of products, sea that made possible imperial exports, conquest of markets. From 1948, the ICI roundel featured more frequently on commodities, transport and so on, the initials I, C, I, set darkly against a background and atop the wavy lines of the nearby ocean, an ocean across which all manner of commodities and materials would travel. The waves were choppy, three peaks, difficult waters, but if the company stayed afloat it would get somewhere.

To be afloat meant steering the boat in ICI's preferred way. In the company magazine of April 1954, a montage of newspaper columns sputtered a double spread of outrage and indignation at suggestions from the Labour Party that the company might be nationalised. The Financial Times, The Daily Mail, The Glasgow Herald, The Daily Herald, The Manchester Guardian, The Daily Telegraph all expressed the ICI directors' point of view. 'ICI reply to Socialists—Tight Monopoly Denied'; 'ICI Answers Back—The Threat of Nationalisation'; 'Case Against State Control of Chemicals'; 'Labour's Plan Attacked—Take-Over Not in Nation's Interests'; 'Quiet 'Mr ICI' will fight Socialist Grab'.[59] The newspaper chorused the findings of 'the quiet man', Dr. Alexander Fleck, whose address to stockholders and employees warned how nationalisation would slow down decisions and make ICI uncompetitive. It would stifle commercial and technical initiative. ICI would lose good quality staff. Nationalisation would damage overseas interests. In July 1954, the same technique of reproducing clippings from newspapers, lending a sense of verity, shared the story of the ICI Plan. 'ICI workers will say "Thanks a million"' stated the Daily Express on 21 May 1954.[60] A profit-sharing plan was launched. Workers were to receive some shares. The Daily Express report was reproduced:

> The bonus will go to all of I.C.I.'s workers in Britain who are over 21 and who will have served two and a half years with the group come June next year. Only in its first year will the scheme cost a million - after that it will grow as more workers qualify.
>
> A fine scheme which the bosses can be proud of. But no doubt the long-haired boys will mumble about it watering down the 'Ordinary capital' and 'reducing the earnings cover'.

Such talk is nonsense. The scheme will mean issuing around 500,000 I.C.I. shares ·a year-against the 140 million that will be floating around.

And by the time ICI has set the £1,000000 cost against profits for tax purposes the scheme will cost it hardly anything.[61]

In June 1954, Deputy Sales Controller J.H. Townsend reported under the title 'Half a Million a Day', for the ICI magazine on the record sales of 1953. Those in the know, he stated, would recognise ICI products in the shops, ICI packeted salt and soda crystals, Dulux and Du-Lite decorative paints, Brushing 'Belco' ranges and 'Lightning' fasteners, 'Luron' fishing casts and Eley-Kynoch shotgun cartridges, and some nylon and 'Terylene' clothes. But most of the products made by ICI were sold to industry. More than 50,000 customers, several thousand different products: cement and plasterboard used in construction were produced at Billingham, carpet with Ardil was made by Nobel Division, glass with soda ash by Alkali Division, a plastic washing bowl of Alkathene, copper tubing, enamel finishes in a washing machine came from Paints Division, dyes for your clothes, corrugated metal, car upholstery in 'Vynide' from Leathercloth Division, refrigerant for ice cream, fertilisers, rubber tyres on vehicles, ICI water treatment chemicals such as 'Afloc', caustic soda for the paper on which the news was written, coal exploded from the ground with Nobel Division explosives and so on (see Fig. 8).[62]

All these, chemicals, substances, fabrics and fibres, would keep the company buoyant. Some of the newer materials began their life at the new site at Wilton. At Wilton, a tall slender chimney stretched two hundred feet into the sky with a pilot light at the top, the flare stack. At Wilton, thick black oil arrived from the Middle East, often passing through the Suez Canal, and some was taken for the olefin products, all those compounds made of hydrogen and carbon whose names end with 'ene', such as Ethylene and Alkathene, Propylene and Butadiene. Ethylene and Propylene were used as chemical feedstocks to fuel vast chemical reactions, but they were also material for other products. Ethylene was made into antifreeze for cars and as a building block to produce polyethylene, PVC and styrene. Propylene found its way into paints and Perspex and plastics such as Polypropylene. Alkathene was used for polythene and Terylene. Butadiene was used for the Butakon range of plastics and artificial rubbers for hoses and shoe soles. Through all this, we became plastic people.

Fig. 8 Alkathene advertisement from the 1950s

Some plastics failed. Others thrived. The company invested more than £2 million in an Ardil production factory, in Scotland, to produce the synthetic material from 1951.[63] The advertising tagline ran 'Happy families with Ardil'.[64] In 1953, the product was showcased at the British Industries Fair.[65] But the factory was closed in 1957. After the war, the price of wool fell and peanuts were becoming more expensive—and short in supplies.[66] The Labour government, under Clement Atlee, attempted to cultivate the Tanganyika region of colonial East Africa into a massive mechanised peanut monoculture, at vast expense, but the scheme failed, not least owing to insufficient understanding of the local ecology.[67] Artificial fibre Terylene, however, flourished.

At Wilton, Terylene was spun, by feeding chips of polymer, into melt heads, and forcing them out under high pressure through spinnerets, which were steel plates perforated with tiny holes. The streams of molten polymer cooled into fine threads, which were then combined into yarn and wound onto bobbins. Processes kept elasticity in the threads, as they twisted together. Terylene sold itself through fashion, through the photographs of models draped in Terylene, there to be looked at. Terylene was a material under research for war purposes, but when it came, it was not for a world of struggle but for one of ease. Terylene negated work—'no ironing' stated in countless advertisements. The pleats are permanent. Terylene never dies, but fashion does, so there would always be the need to buy more clothes made of Terylene. In the ICI magazine of February 1954, the qualities of Terylene are listed: exceptional strength; withstands heating; absorbs very little moisture and has good insulation; chemically very inert; exceptionally durable. Terylene was easy to look after and needed little care and upkeep. 'Terylene' underwear, shirts, dresses—in fact all garments made from the filament yarn—were quickly washed and quickly dried. They did not shrink in the washtub, and, if tight wringing was avoided, they seldom needed ironing. Moths did not eat it. It was tough. But, the consumer was reassured, it could appear dainty in voiles and curtain marquisettes, brocades, satins and velvets, and it was delicate in garments that had to be light, such as summer-weight underwear, shirts, dresses, blouses and ties. It looked good and felt good with 'warmth and pleasantness of handle', and was, in short, alone amongst synthetic fabrics.[68]

Photographs in the magazine, for example, those in the November 1956 edition, in an article titled 'Round the Shops with I.C.I Fibres', showed examples for men and women and children now available in the shops.[69] The clothing had unmodulated colours, a consistency that never faded, on evenly woven, never-creasing, or, alternatively, permanently furrowed fabrics. The heading stated:

> Fibres Division was formed in April 1956. It was a move so long expected that the announcement caused little stir. But today its products are making their influence felt and are now increasingly available in the shops.[70]

Terylene stockings offered 'greater sheerness, more comfort, and wonderful washing qualities' for 9s. 11d. Terylene dresses come in a wide variety of weights and styles for all seasons. Ardil dresses offered warmth and 'perfect comfort'. They could be purchased in the newly built large shopping centres. Pillows were filled and covered with 100% Terylene and were becoming easier to find, at a cost of 35s. to £3. Terylene's fibres tied the whole world together, as it found its way into garments, underwear and lingerie, shirts, ties, blouses and electrical cords, chemical filtration nets, ropes and cords. Terylene net curtains were resistant to sunlight, to the damaging effects of wash and wear. Unlike stretchy nylon nets, Terylene nets were crease-resistant and pleated, for those final years when lives were still led behind synthetic lacy nets. By 1958, according to the April edition of the ICI Magazine, the price had fallen sufficiently that the 'poorest fishermen in India' are now trying out Terylene nets and may decide their long life makes the extra cost worthwhile.[71]

Ulstron came in 1960. Made from polypropylene, it was for prosaic ends, as bristles and pan scrubbers, a competitor to nylon and natural fibres, and able to be made more cheaply. Fabrics multiplied—synthetic voiles, marquisettes, satins and taffetas made to simulate a life led in luxury, on the cheap. These fabrics, moving accessories, introduced new sensations and novel modes of shimmer and drape. And then there was Crimplene, whose patent ICI bought after its invention at the end of the 1950s. Crimplene came to be the material for going-out clothes for working women and men. The colours were bright. It was easy to cut. It did not fray. It held its shape and was dripped dry. It was the stuff of A-line skirts, then miniskirts, of multicoloured men's shirts and slacks. It promised to be clothing for a carefree time of pleasure and leisure.

Crimplene was a heavy texturised cloth of polyester, an augmented Terylene. Its name was a pun. It nodded to the ICI headquarters at Harrogate, for the Crimple Valley was nearby. But crimp also means folding or twining together. Crimplene was usually double-knit—two fabrics twined into one, thick, a real coating for the body. An advertisement from *The Observer*, 25 September 1960, condensed the sense in which these synthetic fibres contracted into themselves all possibilities in the world. They were magic weaves of multiplicity. Crimplene, it stated,

> ... looks pleasantly weighty, yet feels light as a feather. It feels soft, yet can take any amount of wear. It has all the plusses of 'Terylene'.
> Anything made of 'Crimplene' – however haute couture it is – can be washed at home (even in a washing machine) and hung out on the line to dry without losing its shape. No ironing. Sounds too good to be true. But it is true.

What could now be true in the world was not those eternal verities of old, but, instead, lab-made truth. And it took us closer to what we had always desired: a life of ease and plenty lived amongst beauty. The 1960s would usher in—the advert noted in passing—a world of domestic appliances, with their promised labour-savings. Soon, there would be more time to give over to our new beauties.

The ICI magazine from April 1957 showcased an Ideal Homes exhibition kitchen, with Perspex elements, Darvic PVC and nylon.[72] There were also Alkathene housewares displayed in the shop window, brightly coloured polyethylene products. The products, the brand names hinting at substances, cascaded from the factories. Vinyl coated 'Vynair' arrived: 'an entirely new type of upholstery covering', instantly cleanable with a damp cloth, regaining shape quickly too. Vynair 'is a coated cloth with a difference—it breathes!' (see Fig. 9).[1]

And if proof were needed that these vibrant materials promised not only to stimulate the new lives that might be led amongst synthetic products, it was also the case that they themselves possessed a certain liveliness, at least a commodity fetishised liveliness. The ICI magazine from July 1957 gave details of the animated advertisement *The Alkathene Circus*, directed by Etienne Raik, shown in 700 UK cinemas from June to November of that year, in which, as the magazine copy put it: 'gaily coloured cups, saucers, salad-shakers, kitchen brushes – 200 "Alkathene" objects in all – appear to be endowed with a life of their own'.[73]

And what of the lives of those who made them? The workers who clothed and fed the world—and seemed to build a colourful, carefree

---

[1] *The Countryman*, Cyprus. Department of Agriculture, 1958, p. 30.

Fig. 9 Vynair advertisement, 1959

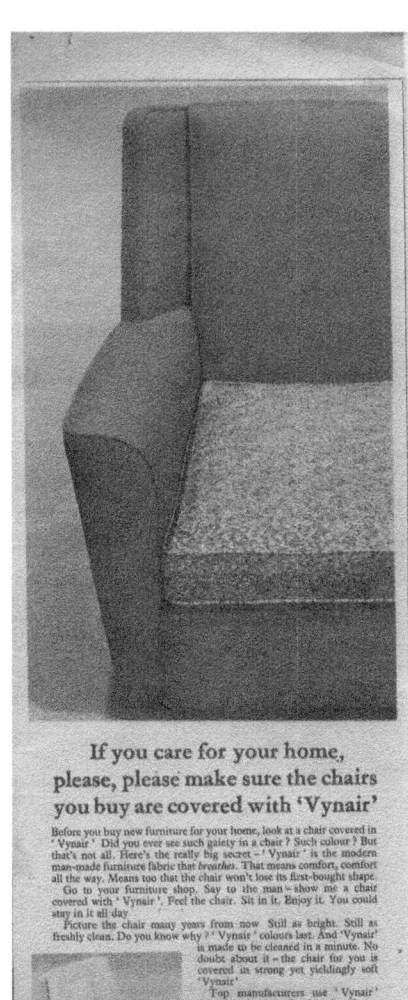

existence for all—were, in reality, in Teesside, as elsewhere, working with nasty substances, including the heavy, dark, thick, cloying, damaging substance of oil. A 1958 company report surveyed the current state of production. Across the company as a whole, alkylamines and derivatives of phenol were 'virtually the only Divisional products that are not derived from petroleum'. The Billingham/Wilton complex made from oil or its derivatives an integrated range of products such as solvents, detergents, plastics and fibres, and was the biggest European centre of petroleum chemical production.[74] But France, Germany, Holland and Italy had plans to expand chemicals from petroleum. New products, KayNitro fertiliser and a new form of Nitro-Chalk, provided free-flowing granular fertilisers, suitable for mechanical handling, to feed British farming, now described as 'the most highly mechanised in the world'.[75] A bold red and blue graph depicted how ICI products served the home market. Textiles were dominant, followed by agricultural products, engineering, chemicals, building trades, motor and aircraft, plastics, wholesalers, mines and quarries, government departments, mineral oil refining.

Nature and the Elements

ICI, like other chemical companies, explored matter and its properties. This matter made history, or changed history, changed the world and what the world was and had within it. The matter extended from the solidity of rock to the abundance of liquid water to the gaseous air and beyond—and it lent itself to certain uses. This matter made certain arrangements possible and played its part in shaping what could be made. Transport systems brought other substances into the area or transported stuff out and across a corporate body so more and more small bits could link together in chains to produce larger things that became ever so necessary to the world, where once they had never existed, or not in that form.

There was something elemental about industrial chemical production. Elemental in the sense of relating to or embodying a process of nature. The factory's blast furnace was like the heat at the centre of the earth, producing reactions, through immense heat bringing about chemical change. In this, it was a replication of nature's work in miniature. ICI's mission was elemental, in that it was to take the elements and make something else of them, to make things that hold within themselves all the elemental power of nature and its capacity for production and destruction. The Billingham factory was created to turn air and water through the use of fire into explosives, to destroy vastly, and fertilisers, to produce exponentially. The process is repeatedly described in the various issues of the company magazine. In May 1954, for example, the lead article was called 'Harvest from the Sky'—a film under the same name was made by ICI Billingham Film Unit around the same time.

> To most of us ammonia is a cloudy, pungent liquid sold in small bottles for rather vague household purposes that include the revival of those on the point of fainting.
>
> In fact it is a very valuable chemical raw material. At Billingham alone it is used in the manufacture of all three of the fertilizers - sulphate of ammonia, 'Nitro-chalk', and concentrated complete fertilizers - which make up such a large part of the Division's output; and it is equally important as a raw material in the processes producing nitric acid, ammonium nitrate (some of which is used for blasting explosives), urea for plastics and last, but by no means least, the nylon produced by the Billingham factory of Dyestuffs Division.[76]

This crucial stuff, it reported, was made simply of wind and rain. Another element was added to the image of the factory: fire. At Billingham, coke was made. Drivers rammed it from the ovens and the red-hot lumps cascaded into a truck, to be quenched with water. 'Hissing, billowing clouds of steam roll skywards—a delight to the spectator'.

Inside tall steel towers, in the making of ammonia, reactions were triggered by catalysts as nitrogen was seized from the air and hydrogen was taken from water. In the process, large quantities of carbon dioxide were removed from the mix—this was used further for making ammonium sulphate, but some was solidified and turned into dry ice—marketed by ICI as Drikold—a coolant, a solid frozen mass, an artificial ice, made into something usable, controllable—a coldness that can be conjured at will to keep fruit and flowers fresh-seeming, to allow fish to travel to distant markets, to make the artificial snow needed for artificial ski slopes. Drikold was made in Billingham and, in its early days, it was a product used for cloud seeding, for the control of the elements and for the control of weather. The clouds were seeded over Teesside with Drikold for the first time ever. New clouds, synthetic clouds, were made here in 1931 and caught on film.

The towers made clouds of gas and steam. These clouds made more clouds, or smogs. Smoke and sulphur dioxide, which had been derived from burning, for example, make condensation nuclei, or cloud seeds and these encouraged condensation at ground level, which resulted in opaque mists. But it was also the case that clouds were seeded in Billingham, that is to say, deliberate and real ones of rain. Or more accurately, the clouds that hung high above the land were prodded and supercooled and made to release their watery contents. A film made by the Billingham ICI Film Unit in 1931, titled *Weather Experiments*, documented the process of seeding clouds with Drikold. The film began with a radio announcement of the weather forecast. Viewers saw a meteorological officer listening in to the forecast from his living room. He switched off the radio so that he might convey to us the value of weather forecasts. He told of how wind speeds were measured and wind direction monitored. Inflatable balloons launched. Weather stations supervised. All this must be known so that the country could keep running what needed to be run. In this film world, men smoked as they read charts. And teletype data flowed in. Shown is how nature worked away and instruments measured it. But revealed here, in this film, too, was how the weather could be manipulated, the clouds formed by human ingenuity. Cylinders of Drikold moved along a conveyor belt and were packaged into bags. This Drikold was placed on aeroplanes that flew above the clouds. In this film, 300lbs of Drikold was shown releasing from the rear of a plane, while inside the navigator recorded the atmospheric conditions and observed the location of the penetrated cloud. Its watery load streamed down the windows of the aircraft. Rain had been made. ICI were weather modifiers. ICI could rule the heavens.[77]

The experiments into weather manipulation continued, or resumed, becoming part of military training, for the advent of sudden rainfall might bog down an enemy's movements or prevent the crossing of a river. A shower might clear the fog on an airfield or divert a storm system from the path it was on. Documents can be found that speculate on rainmaking possessing the potential 'to explode an atomic weapon in a seeded storm system or cloud', one made in just the required place. The thought continues: 'This would produce a far wider area of radioactive contamination than in a normal atomic explosion'. In August 1949, ICI worked with the Royal Air Force to generate rain through seeding clouds. A Vickers Wellington TIO training aircraft, which was based at RAF Middleton St. George, outside Darlington, flew three sorties with a crew to drop Drikold over the North East and the North Sea. HQ Flying Training Command observers were able to report on 'a good shower' at the cloud's base, on one flight, but, on the other two days, there was only a light drizzle that evaporated in the air. The experiments using Billingham chemicals continued and, on 15 August 1952, they may—some suggest—have caused a flash flood in Devon, in the village of Lynmouth.[78] Alongside some hints in official documents, unofficial reports insist that aeroplanes were circling the area before the catastrophic downpour. Whose stories? Whose histories? Thirty-five people died. Water gushed down from the moor into the village, sweeping away buildings and bridges. The Squadron leader joked that the efforts to make rain, called officially Operation Cumulus, was known to them as Operation Witch Doctor. From Billingham, a hell could be whipped up far down in the South of England. And a storm from precipitated clouds, heavy with dense particles that have been shot into its body, might drop its water load some hundreds of miles away. History is meddled with by nature. Nature is subjected to covert histories.

And turning attention away from the skies above Teesside, what did the landscape's interior offer up? In an article in the ICI magazine from October 1951, a vision of 'the Billingham Underworld' was painted. Eight hundred feet beneath Billingham existed a maze of tunnels, divided by 5,000 broad pillars of rock and abuzz with electric trains and diesel wagons, criss-crossing 200 miles of well-lit roads arcing up and downhill in pursuit of the seams. The mine was a square mile, a mile squared of blasted rock, whose relation to the abstract transactions of the square mile of the City of London, where finance was concentrated, was certain, but difficult to grasp. To reach this underground city, each miner drew a numbered brass token from the clerk at the pithead office and handed it to the banksman as he entered the cage. Here, beneath the ground, was a hell of production, with exploding gelignite, noise, smells. The magazine article described the scene in mythic terms: when an explosive shot was fired down there in the guts of the earth it was as if a furious giant slammed the Gates of Heaven, for a vicious sheet of sound whipped through each miner's body. These works were for the mining of anhydrite, 19,000 tons a week of which was hauled to the surface to feed the plant. Anhydrite means 'without water', and so, next to the sea and river, sat a mass of rock that lay where the sea once was, now evaporated. This calcium sulphate or dehydrated gypsum gave up its calcium for cement and the fertiliser Nitro-Chalk. Its sulphate went into sulphate of ammonia, another fertiliser, and into sulphuric acid, or hydrogen sulphate. From beneath itself, the factories' main products were extracted. And the world should have smiled, for the report noted, cement was in short supply and fertilisers were essential for growing grass to feed the cows and replace the animal food which we could not afford. And sulphuric acid was even more valuable in the context of a world sulphur shortage.[79]

Underground too was life reproducing itself. The February 1954 edition of ICI magazine reported from Billingham Division on the opening of a works canteen.

> The cream distemper, red-tiled floor and concealed lighting of the new canteen at Billingham might belong to any modern canteen. Only by the rough-hewn walls and the miners' helmets on the men's heads can you tell that this canteen is 800ft. underground in the anhydrite mine.
>
> There is nothing like it anywhere else in Britain - this place where the men who work in the underground workshops and service the mine vehicles and equipment can have their meals, and where all who work in the mine can see films on safety and other subjects.
>
> One of the galleries in a disused working has been used to make this room - some 18 ft. in width and about three times that in length. In order to close it, openings in the gallery have been bricked up, but most of the walls are left roughhewn just as they were when they ceased to be worked.
>
> The room is on two levels, the upper one, equipped with chairs and tables, being reached by a short flight of steps. The lower level is where the men can wash the things they use for meals at sinks along one wall. High up on the end wall of this lower level is the cinema screen, and the projector will be installed in the canteen on the upper level.
>
> Lighting is through two rows of louvres in a false ceiling in the roof. An unusual feature of this ceiling is that above it and along its whole length are two walkways so that the rock roof can be examined regularly, as required by mining regulations. Other amenities include washplaces and toilets. The whole project was planned by the mine engineering staff and designed by Chief Engineer's Department, and work began three years ago. It has been carried out by the Services Section of Engineering Works and the men in the mine, with an outside firm to lay the floor tiles and put in the false ceiling.[80]

Amongst this hellfire of production, men must live and keep themselves alive. How old might all the structures of labour seem, mythically ancient, when, trapped in a cave, like busy dwarves in a fairy tale, the workers are held below ground?

But the food would be abundant. The issue from March 1954 considered the question of food provision. Kevin FitzGerald, from Central Agricultural Control, was the writer of 'Nearly as Much Meat As You Like'. It was an article in praise of fertilisers. It reflected on how all people around the age of twenty would have experienced only food shortages and difficulties in their lives, rationing in and after the war for the most part, ersatz and poor-quality food or small amounts. Sausages exploded or melted in the pan, blue meat with blue marks indicating various countries of origin, horse meat, donkey meat and mule meat. ICI fertilisers had changed all of that and now. By the mid-1950s, an epoch of abundance had begun. Utopia had arrived. In utopian landscapes, such as Cockaigne, Schlaraffenland or the Land of Milk and Honey, foodstuffs come with ease, sometimes with knives and forks already plunged. In Breughel's *Luilekkerland*, his lazy luscious land, the workers are sated and lie flat. The houses are roofed with pies. The wine has flowed, without effort, straight into the mouth. A pig ambles by, the carvery knife inserted into its flank for easy pickings. Might ICI make real such longings? It could command the skies and make the weather like a God.

A drought in the North East in the autumn of 1955 raised the issue of bringing rain to parched land—for the problem had returned home. Billingham's ICI factory was threatened with closure and management required aircraft to spread its own chemicals in the skies above, so that they might go on making more clouds.[81]

In the company magazine from December 1959, G.E. Stewart, reported, in 'Modern Mechanised Mining', on the full-automation of the anhydrite mine.[82] This was a sign from the future. At first the harbingers seemed good. Full automation would bring about an increase in leisure. This leisure would be enjoyed in homes that are bright and beautiful and fully synthetic. They would barely need to be cleaned. Propathene, ICI's brand name for Polypropylene, could make everything in its image, from children's toys to cups and saucers, to hospital trays, food wraps and washing machines and suitcases. In April/May 1963, the magazine listed the twenty plastics in commercial production, the most important being acrylics, nylons, fluorine-containing plastics, Polypropylene, silicones, polyurethanes, polyformaldehydes, melamines, and reinforced plastics based on polyesters or on epoxies. Each was, according to Cyril Child, in his magazine article 'Trends in Plastics', available in many grades and formulations, and some in different physical forms such as liquids, lumpy solids, powders, granules, sheets, films, rods and tubes. Some are used as virtually pure materials, some have many different materials compounded with them to modify their properties (including their prices) – additives such as plasticisers, extenders, fillers, heat stabilisers, light stabilisers, antistatic agents, antioxidants and pigments. Some grades show subtle differences of physical forms such as different particle shapes or particle size distribution.[83]

The front and back cover of ICI magazine from June/July 1966 presented views of No. 4 Olefine Plant, operated by the Heavy Organic Chemical Division at Wilton. The front of the magazine showed it by day. The sky was blue with wispy clouds, some of which might have leaked from the tall metallic towers, silvery against the sky, except for one blue shaft. The scene was perceived through a black metal structure that seemed oily and heavy. The towers rose like slender poles. This was the face of modern industry. The image on the back cover was taken at night. The sky was yellowy orange with the setting sun and the clouds, still perceptible in their last moments of visibility, were deep grey, like smoke balled from an industrial disaster. Now the towers, photographed from closer in, appeared black and gloomy. The complex structures, a filigree around their edges, stood out in this light. The lacing was the outline of service access ladders, necessary human infrastructure to make these things work. The caption noted that it 'is in the process of being commissioned'. The vast quantities of ethylene it will work with will expand plastics production, notably Alkathene and vinyl acetate. It will make propylene and butadiene too.

Futurology

ICI's company magazine, as it developed into the 1950s and 1960s, promoted a sense of colour, a visual boldness, that was promulgated in its consumer products. ICI represented a world that was beautiful, from its factory interiors and exteriors, be they in Billingham or Brazil, Singapore or Welwyn Garden City, a world that was vast and where trade flowed easily across borders. It was a world beautified by synthetics—gorgeous fabrics and wallpapers and coatings—and synthetics also produced abundance, fertility and chemicals that remade lives. Had DuPont, the US chemical company, not taken the slogan already, ICI would have claimed to deliver 'Better Living Through Chemistry'. The roundel with the company initials, ICI, and wiggly lines to represent waves, introduced in the early days of the company was adapted. In 1969, Design Research Unit modernised the logo, plumping for then fashionable orange as a dominant colour, along with its black or sometimes white lettering and wavy lines representing the sea.

In the 1960s, research at ICI looked into aerosol substances and mechanisms. It was observed to be a growth area, with some 70 million aerosols consumed during 1963 by what were termed 'an ever more aerosol-conditioned public'.[84] The report had a photograph of a wide array of entries in an aerosol package competition, held at ICI House.

> Here, a bewildering display of every type, size and contour of aerosol was arranged in groups according to their functions, from the largest and most utilitarian of insecticides to the smallest and most elegant of scent-sprays, diminutive enough to go in a lady's handbag.[85]

It was observed that, by 1970, annual production of aerosols in Britain was expected to reach 200 million. This modification of air was for many uses: insecticides, perfumery, cosmetics, polishes, hair lacquers, air freshening, mothproofing, lacquering, painting, pharmaceutical, veterinary medicine or horticulture. The claim appeared: 'the aerosol has become, and is destined to become still further, an indispensable adjunct – if not indeed a positive symbol of what has significantly been called the Press-Button Age'.[86] The Press-Button Age: name for a time of ease, of lazy transformation of environments, of a synthetic atmosphere in a can to be deployed at will, anywhere, anytime. If reference is to fingers on buttons, might mention here also be made of the fact that in the 1940s research into tube alloys at Billingham, under Maurice Hodgson, and the work of ICI engineers, such as Christopher Hinton, who engineered piles and separation processes, in the Industrial Group of Britain's Department of Atomic Energy, contributed towards the development of the atomic bomb, whose detonation was, apparently, just a push-button away.[87] But this did not play into the time of its happening, emerging only in hints and rumours in the post-war years.

Future behaviours could be moulded. In May/June 1968, the ICI magazine, in its stylishly lower-case article 'at home with colour', reported that householders were moving away from only redecorating their homes when they really had to. But consumers needed to be inducted into the new world of bold hues now available. The possibilities of bright colours and patterns, lightweight materials—Dulux's paint range, Vymura vinyl wallcovering, WalFlair wallpaper—needed explanation, which would stimulate demand, as well as the desire to redo the decor whenever a householder felt like a change. A colour schemer had been devised—designed by Jack Widgery and John Lupton of the Colour Advisory Department at Slough—to orient people in this bright new world:

> The first of its kind, the Colour Schemer looks rather like a wallpaper pattern book and contains 68 room schemes divided into eight 'mood' sections, e.g. sunny, dramatic, romantic. Each section contains both coloured illustrations and colour scheme suggestions combining paint and wallpaper.[88]

By May 1969, ICI was convinced it could play a role in 'Making Tomorrow Happen'. This was an article's title, written in something like Leo Maggs' Westminster typeface, as designed for machine reading by a simple magnetic reader capable of automatic character recognition. The contribution speculated on a future into which ICI should fit, or which it should mould. It provided a vision of increasing consumer-purchasing power, fewer hours devoted to work, a higher quality of life and more automation:

> Nor should we forget that by 2000 AD nearly 16 per cent of the population (11 million) with leisure will be over the age of 60. Lightweight clothing, lightweight household objects, gardening tools, etc., will be very much in demand.[89]

But the article's author was not ignorant of the looming negative possibilities. In the 1960s, increasing attention was paid by ICI, and the chemical industry more broadly, to the role of computing. The author of 'Making Tomorrow Happen', Norman MacLeod, a chemical engineer, stated:

> There is a negative side. Although the growing power of the computer to store knowledge about many aspects of an individual's life and history is recognised as being in some cases beneficial, misuse of this power would be an unwarranted interference with the privacy of the individual. For my own part, I believe that society must be prepared to resist such misuse.

Another prospect also repels. The use of new drugs, which might be developed and might have the effect of enabling people's minds to be conditioned, would have to be rigidly controlled. We take measures to control certain drugs which, on balance, have a bad effect on people – such as heroin. We may have to protect society from the misuse of new technological developments in the same way.[90]

Drugs and technology were twin aspects that in that moment were coming together in new ways through the counter cultural movement of the Whole Earth Catalog.[91]

There was a less visible, but more persistent, future-oriented danger. In the December 1958 edition of the ICI magazine, a feature on 'The World of Toys' crowed about the competitive advantage that toys made of plastic from ICI now had in the world, beating the historically rivalrous German toy production.

> The British toy industry is today composed of a large number of highly competitive small firms and is second to none, having assumed the lead held pre-war by Germany. The modern range of plastic toys owes a lot to I.C.I. plastics and in particular to the I.C.I. discovery of polythene.[92]

Supplied by Hamleys, the prominent London toy store, they rendered in miniature a world of animals, vehicles and war. A foam rubber guardsman never fainted on parade and could take up any position for 9s. 6d. Polythene beads could be pulled apart and snapped together again by a baby. There were launching rockets of nylon, polythene tank transporters carrying jeeps, a Red Star rifle, Night Rider pistol and Space missile gun, which shot suction darts that stuck to their objective. Fads were reflected: 'Cowboy "Swoppets" are new this year and selling fast. Made of polythene, they take to bits on the poppet system and can be put together with heads, arms, legs, and even pistols, on different bodies'.[93] A build-it-yourself polythene double-masted schooner, called the 'Black Falcon', was showcased alongside a build-it-yourself B.O.A.C. Britannia aeroplane. Cheap, disposable toys, made for just as long as a childhood lasted, do not disappear with its ending, but persisted, our plastic siblings who never died, just faded slowly.

'ARCTIC JUNKY—Plastic Sugar Puffs toy from 1958 washes up in the Arctic 59 YEARS later to give a chilling warning about plastics in our ocean' states the newspaper headline in 2017. A tiny replica of the RMS Mauretania bobbed its way on the oceans 1500 miles from Britain, from its cereal packet existence, via the gutter and dump, set adrift on a sea for which it was never destined to come ashore on Jan Mayen Island.[94] 'Ahoy! A Free Toy!' was the advertising tagline from Quaker, and millions of children rummaged in the packets to find this plastic booty. The material body of this consumption-stimulating giveaway, however insignificant and throwaway its design, persisted, decaying only gradually, its colour changing from pink to cream, one funnel going missing. It probably deteriorated slower than that of the huge ships and tankers that passed over it. It decayed slower than the original of which it was a copy—a Cunard-White Star transatlantic ocean liner, built in Birkenhead and launched in 1938. It was scrapped at Thos. W. Ward's shipbreaking yard in Fife in the mid-1960s.

Over time, these plastic things leaked their being into the water. Just as did the synthetic garments of a wonder-stuff such as Terylene, its plastic fabric fibres formerly sold on the promise it would hold its own for wash after wash, and yet all the while it unhurriedly gave up its substance, its microfibres escaping into water and into the world. The Terylene garments slowly deteriorated, as they were subjected to slight shifts in humidity, the impact of oxygen or heat. They gathered dust, more dust than natural fibres, for they were electrostatic. Greenhouse gases, like nitrous oxide, and volatile organic compounds are emitted by synthetic fabrics, as they age, as they are destroyed and as they sit in landfill. When the synthetic garments were washed, microfibres escaped into the environment. These things break up. These plastics bleed, trickle and shatter quietly to return as deposits in the ground, sea and bodies. This too was a truth of the items—for, as much as they were durable, they were also cheap and made to be thrown away. Some became plastiglomerates, or pyroplastic lumps, fragments of burnt plastic, which, if weathered enough, erode down in the swash zone into clasts that are geogenic. They find a way back to the places from which they derived (before they were they and a thousand complex processes divided and recombined them)—back to the sea whence came crude oil, back into the rocks from where natural gas was extracted, or into the ground, concentrated in landfill, dispersed in environmental pollution, to greet the stocks of coal now almost exhausted. As the plastics dispersed, so too did memory of each item's promise of better lives lived with them. Their bold colours, their resistance to the environment, to water, wind and sun, their capacity to form anything and everything, by tricking nature, or outwitting it—and holding all this for us forever—proved to be a lie. What contradictions are held in this cartoon substance of plastics, once morphed into infinite possibility and endless invention, promising to be there forever—durable, consistent, impervious—while made to be throwaway, un-insistent, happy enough to stimulate economies through desire-driven replaceability? They were bought to be durable. They were bought to be thrown away. They were thrown away, in time, and they remained durable, or durable enough to never disappear. But all in the wrong way. That they persisted, while giving off small parts of themselves into the environment, became a problem in waiting.

How did people live amongst this chemical concern, beneath and above the ground? What was it like to live in a landscape that made or remade worlds, synthesised environments, took the air and water and made it bright or pliable or hard and scratchy? How did the air smell? Could the gushing water be heard? What sounds escaped? Did the way in which the time codes of nature were cracked, the ways in which the effects of time's *longue durée* were arrested by dry ice or the ways in which processes that nature carried out slowly were speeded up, under heat and pressure, feed into dreams and thoughts and self-image? A 1965 film, by an amateur filmmaker from the Cleveland Cine Club, called *If I film at 2fps…*, depicted a speeded-up race through the streets of Middlesbrough to the north bank of the Tees, in order to reach the factory in time to clock in.[95] Life accelerated, under the power of petrol and the force of factory time.

The future was a horizon for business, but the pressures of the present-day made themselves felt in a 1970 report on Staff Representation. This was ICI's response to the major political parties arguing for responsible collective bargaining over wages and conditions. Management techniques, the report stated, were now more sophisticated. White-collar workers wanted greater representation. White-collar unions were growing and thus the image of the trade union movement was changing. The report asked if unionisation for ICI staff, at a low level at this point, was inevitable. Could it be offset by a Staff Association? The report conceded 'The subject of Trace Unions is highly emotive but the influence of the Unions in the white-collar sector of industry is growing rapidly and it is important to consider the issues as objectively as possible'. The directors feared inter-Union disputes around recognition and a chaotic situation developing. What form would representation take? Staff were considered a third term between Company and Unions: 'Both the Company and the Unions are influencing the course of events in ICI and it is in the best interests of staff, that they too, exert their influence by taking prompt concerted action'. Who were the workers? Who were the staff? What did they want? What did they get?

Notes

1. Rob Perrée, *Bakelite: The Material of a Thousand Uses: Based on the Becht Collection*, Cadre Snoeck-Ducaju, Ghent, 1996.
2. *Plastics and Molded Products*, Vol. 4 (1928): 231.
3. Up until the 1970s, it was written Tees-Side. I use Teesside. Nowadays, the preferred term of officials to name the area incorporating Middlesbrough, Redcar and Cleveland, Stockton, Hartlepool and Darlington is Tees valley. The area including the four boroughs of Hartlepool, Stockton-on-Tees, Middlesbrough and Langbaurgh-on-Tees was, from 1972 to 1996, called Cleveland. Sometimes that name appears in what follows.
4. I use ICI to refer to the company. Some texts use I.C.I., and I occasionally retain that format in quotations.
5. Charles Cullimore, *The Last Days of Empire and the Worlds of Business and Diplomacy: An Insider's Account*, Pen & Sword History, Yorkshire, 2021, p. 122.
6. Jon Warren, *Industrial Teesside, Lives and Legacies*, Palgrave Macmillan, Basingstoke, 2018.
7. Ray Hudson, 'Rethinking Change in Old Industrial Regions: Reflecting on the Experiences of North East England', *Environment and Planning A: Economy and Space*, Vol. 37 (4): 581–96.
8. George Head, *A Home Tour Through the Manufacturing Districts of England, in the Summer of 1835*, John Murray, London, 1836, p. 312.
9. George Head, *A Home Tour Through the Manufacturing Districts of England, in the Summer of 1835*, John Murray, London, 1836, p. 313.
10. George Head, *A Home Tour Through the Manufacturing Districts of England, in the Summer of 1835*, John Murray, London, 1836, p. 314.
11. Minoru Yasumoto, *The Rise of a Victorian Ironopolis: Middlesbrough and Regional Industrialization*, Boydell Press, Suffolk, p. 10.
12. Barry Doyle, *A History of Hospitals in Middlesbrough*, South Tees Hospitals NHS Trust, Middlesbrough, 2000, p. 5.
13. John K. Harrison, *John Gjers: Ironmaster, Ayresome Ironworks Middlesbrough*, De Archaelogische Pers, 1982, p. 10.
14. Minoru Yasumoto, *The Rise of a Victorian Ironopolis: Middlesbrough and Regional Industrialization*, Boydell Press, Suffolk, pp. 40–1.
15. One example: Arthur Newsholme and W.W.E. Fletcher, 'Report to the Local Government Board upon the Sanitary Circumstances and Sanitary

Administration of the County Borough of Middlesbrough, with Special Reference to the Persistently High General Death Rate and Infantile Mortality, and Their Causes', London, HMSO, 1910.
16. See Sven Beckert, *Empire of Cotton: A Global History*, Knopf Publishing Group, New York, 2014.
17. British Committee on Industry and Trade Survey of Industry, Part III, *Survey of Textile Industries: Cotton, Wool, Artificial Silk*, HM Stationery Office, London, 1928, p. 282.
18. Walter Greenwood, *Love on the Dole*, Doubleday, Doran and Co, New York, 1934, p. 56.
19. ICI Magazine, July 1933, p. 13.
20. ICI Magazine, July 1933, pp. 17–18.
21. ICI Magazine, July 1933, p. 19.
22. ICI Magazine, July 1933, pp. 91–2.
23. 'A short history of Billingham up to the time of the formation of ICI', 28 July 1939. Draft typescript, 'representing an amplified version of Mr RT Trotter's original shorter account', by D Gardner. Archived at Institute of Mechanical Engineers: Sir Frank Ewart Smith Papers, 1922–1995: FES/3/1/20.
24. See Kohei Saito, *Karl Marx's Ecosocialism: Capital, Nature, and the Unfinished Critique of Political Economy*, Monthly Review Press, New York, 2017.
25. G. F. Whitby, 'Sir Frank Ewart Smith', *Biographical Memoirs of Fellows of the Royal Society*, Vol. 42 (1996): 420–31: p. 422.
26. 'Nitrogen Fixation. (Billingham Factory)'. HC Deb 7 April 1925, recorded in Hansard.
27. Peter Allen, 'Francis Arthur Freeth', *Biographical Memoirs of Fellows of the Royal Society*, Vol. 22 (1976): 110.
28. G. F. Whitby, 'Sir Frank Ewart Smith', *Biographical Memoirs of Fellows of the Royal Society*, Vol. 42 (1996): 420–31: p.422.
29. 'Nitrogen Fixation. (Billingham Factory)'. HC Deb 07 April 1925, recorded in Hansard.
30. 'Nitrogen Fixation. (Billingham Factory)'. HC Deb 07 April 1925, recorded in Hansard.
31. 'Gland Packings for High Pressure Gas Circulators and Compressors', 3 October 1923. Archived at Institute of Mechanical Engineers: Sir Frank Ewart Smith Papers, 1922–1995: FES/3/1/10.
32. 'A short history of Billingham up to the time of the formation of ICI', 28 July 1939. Archived at Institute of Mechanical Engineers: Sir Frank Ewart Smith Papers, 1922–1995: FES/3/1/20, p. 9.
33. 'History of the Process Department (Nitrogen Division)', 15 April 1937. Special typed script report by FM Ray. With an annotated note by Smith

'Not very balanced, complete or accurate, 11/11/80'. Archived at Institute of Mechanical Engineers: Sir Frank Ewart Smith Papers, 1922–1995: FES/3/1/19.
34. 'The Most Economical Method of Bagging Sulphate', 6 February 1926. Typed script report by Smith, being Laboratories technical report no.1263. Archived at Institute of Mechanical Engineers: Sir Frank Ewart Smith Papers, 1922–1995: FES1/1/15.
35. Walter Benjamin, 'Die Waffen von morgen', *Gesammelte Schriften*, Vol. 5; Rolf Tiedemann and Hermann Schweppenhäuser (eds.), Suhrkamp, Frankfurt/Main, 1972–1991, pp. 474–5.
36. Walter Benjamin, 'Surrealism: the Last Snapshot of the European Intelligentsia', *Selected Writings, 1927-1930*, Vol 2.1 (translated by Edmond Jephcott), Belknap Press, HUP, Cambridge, MA, 2005, p. 217.
37. 'Electrical Review', 1930. Archived at Institute of Mechanical Engineers: Sir Frank Ewart Smith Papers, 1922–1995: FES/3/1/18.
38. 'Electrical Review', 1930. Archived at Institute of Mechanical Engineers: Sir Frank Ewart Smith Papers, 1922–1995: FES/3/1/18, p. 1.
39. 'A short history of Billingham up to the time of the formation of ICI', 28 July 1939. Archived at Institute of Mechanical Engineers: Sir Frank Ewart Smith Papers, 1922–1995: FES/3/1/20.
40. George Orwell, *The Road to Wigan Pier*, Penguin, Harmondsworth, 2021, p. 76.
41. George Orwell, *The Road to Wigan Pier*, Penguin, Harmondsworth, 2021, p. 74.
42. George Orwell, *The Road to Wigan Pier*, Penguin, Harmondsworth, 2021, p. 74.
43. ICI magazine, October 1957, p. 355.
44. 'Chemistry at your service: an exhibition of some achievements of British chemical research at the Tea Centre', 22 Lower Regent Street, London, S.W.1, 5–28 June 1946. Issued by Imperial Chemical Industries, ltd.
45. 'Chemistry at your service: an exhibition of some achievements of British chemical research at the Tea Centre', p. 8.
46. 'Chemistry at your service: an exhibition of some achievements of British chemical research at the Tea Centre', p. 9.
47. 'Chemistry at your service: an exhibition of some achievements of British chemical research at the Tea Centre', p. 9.
48. 'Chemistry at your service: an exhibition of some achievements of British chemical research at the Tea Centre', p. 9.
49. 'Chemistry at your service: an exhibition of some achievements of British chemical research at the Tea Centre', p. 10.
50. See Marie Stenton, Veronika Kapsali, Richard S. Blackburn, and Joseph A. Houghton, 'From Clothing Rations to Fast Fashion: Utilising Regenerated Protein Fibres to Alleviate Pressures on Mass Production', *Energies*, Vol. 14 (18) (2021): 5654.

51. 'Chemistry at your service: an exhibition of some achievements of British chemical research at the Tea Centre', 22 Lower Regent Street, London, S.W.1, 5–28 June 1946, p. 10.
52. 'Chemistry at your service: an exhibition of some achievements of British chemical research at the Tea Centre', p. 11.
53. 'Chemistry at your service: an exhibition of some achievements of British chemical research at the Tea Centre', p. 11.
54. ICI magazine, March 1950, p. 71.
55. Discussed in Robert Leeson (ed.), *A. W. H. Phillips: Collected Works in Contemporary Perspective*, Cambridge University Press, Cambridge, 2011.
56. ICI Magazine, March 1950, p. 71.
57. G. Pepler and P.W. Macfarlane, *The North East Development Area Outline Plan*, HMSO, London, 1949, p. 100.
58. Cited in Huw Beynon, et al., *A Place Called Teesside: A Locality in a Global Economy*, Edinburgh University Press, Edinburgh, 1994, p. 64.
59. ICI Magazine, April 1954, pp. 106–7.
60. ICI Magazine, July 1954, p. 199.
61. ICI Magazine, July 1954, p. 199.
62. ICI Magazine, June 1954, p. 162.
63. See Marie Stenton, Joseph A. Houghton, Veronika Kapsali, and Richard S. Blackburn, 'The Potential for Regenerated Protein Fibres within a Circular Economy: Lessons from the Past Can Inform Sustainable Innovation in the Textiles Industry', *Sustainability*, Vol. 13 (4) (2021): 2328.
64. Kersten T. Hall, *The Man in the Monkeynut Coat: William Astbury and the Forgotten Road to the Double-Helix*, OUP Oxford, Oxford, UK, 2014, p. 186.
65. Hannah Auerbach George, 'Peanuts in the archive: Imperial Chemical Industries, Tibor Reich and the British Industries Fair', V&A Blog, 16 December 2022: https://www.vam.ac.uk/blog/museum-life/peanuts-in-the-archive-imperial-chemical-industries-tibor-reich-and-the-british-industries-fair.
66. See Marie Stenton, Veronika Kapsali, Richard S. Blackburn, and Joseph A. Houghton, 'From Clothing Rations to Fast Fashion: Utilising Regenerated Protein Fibres to Alleviate Pressures on Mass Production', *Energies*, Vol. 14 (18) (2021): 5654.
67. Stefan Esselborn, 'Environment, Memory, and the Groundnut Scheme: Britain's Largest Colonial Agricultural Development Project and Its Global Legacy', *Global Environment*, Vol. 11 (2013): 58–93.
68. ICI Magazine, February 1954, pp. 48–51.
69. ICI Magazine, November 1956, pp. 338–9.
70. ICI Magazine, November 1956, p. 338.
71. ICI Magazine, April 1958, p. 121.
72. ICI Magazine, April 1957, pp. 132–3.

73. ICI Magazine, July 1957, p. 246.
74. 'Review for 1958', Archived at Institute of Mechanical Engineers: Sir Frank Ewart Smith Papers, 1922–1995: FES/3/1/22, p. 6.
75. 'Review for 1958', Archived at Institute of Mechanical Engineers: Sir Frank Ewart Smith Papers, 1922–1995: FES/3/1/22, p. 8.
76. ICI Magazine, May 1954, p. 130.
77. There is a comprehensive collection of materials drawing on archived items relating to ICI and RAF cloud seeding activities at https://www.whatdotheyknow.com/request/89323/response/218495/attach/4/CloudSeeding.pdf.
78. See, for example, John Vidal and Helen Weinstein, 'RAF Rainmakers Caused 1952 Flood', *The Guardian*, 30 August 2001.
79. ICI Magazine, October 1951, pp. 290–5.
80. ICI Magazine, February 1954, pp. 55–6.
81. Traces of these activities are in the Parliamentary record, Hansard: Roy Mason, 'Rain-Making Experiments (Cloud Seeding)', HC Deb 07 November 1955 vol 545 c161W.
82. ICI Magazine, December 1959, pp. 326–31.
83. ICI Magazine, April/May 1963, p. 44.
84. ICI Magazine, December 1963/January 1964, p. 187.
85. ICI Magazine, December 1963/January 1964, p. 186.
86. ICI Magazine, December 1963/January 1964, p. 187.
87. See Sean F. Johnston, 'Security and the Shaping of Identity for Nuclear Specialists', *History and Technology*, Vol. 27 (2) (June 2011): 123–53.
88. ICI Magazine, May/June 1968, p. 91.
89. ICI Magazine, May 1969, p. 102.
90. ICI Magazine, May 1969, p. 102.
91. See Fred Turner, *From Counterculture to Cyberculture: Stewart Brand, the Whole Earth Network, and the Rise of Digital Utopianism*, University of Chicago Press, Chicago, 2006. See also Diedrich Diederichsen and Anselm Franke (eds.), *The Whole Earth: California and the Disappearance of the Outside*, Sternberg Press, New York, 2013.
92. ICI Magazine, December 1958, pp. 412–7.
93. ICI Magazine, December 1958, p. 417.
94. David Jones, 'What this toy from a 1958 packet of Sugar Puffs tells us about the catastrophe caused by plastic: Found in the Arctic 60 years on, it had travelled 1,500 miles from a British breakfast table', *Daily Mail* online, 22 December 2017.
95. *If I film at 2fps...*, Cleveland Cine Club, 1965: https://player.bfi.org.uk/free/film/watch-if-i-film-at-2fps-1965-online.

The Highpoints and the Low Ones: In, Over, Around and Under the Chemical Factory

Abstract This chapter explores labour conditions and living conditions in and around the factories of ICI. Drawing on resources such as company proposals, a Mass Observation survey and work by industrial sociologists, Huw Beynon and Theo Nichols, it asks what was it like to work for ICI in the 1930s through to the 1970s? It explores how the company contributed to public health and how institutions of welfare arose in its environment as part of a quest for modernisation. The use of plans—e.g. urban plans, work demarcation and categorization—to bring rationality to the commotion of industrial expansion is highlighted. The role of the accident attendant in industrial work—especially chemical work, is considered. Around the factories and plants too is nature, and it is affected by and affects what happens inside the factory. Something, in the form of pollution, seeps out from industry—and it needs to be accounted for. A case study of pollution in Teesside in the 1960s is explored here, through its tabling as a parliamentary concern.

Keywords Crimplene · Guano · Industrial pollution · Process work · PVC · Shift system · Staff grading · Technical training · Teesside plan · Terylene

Who is Making Those Clouds?

The factories soared above the horizon. They poked into the clouds, which hung on them on dull days. They made their own clouds and fog too. Brown and grey clouds. These mingled with the clouds in the sky and with the fog on the ground. Fog and clouds of steam and smoke imprinted themselves onto countless photographs. For as long as the factories existed, they made the air thick. A photograph of the Ammonium Sulphate Towers in Stockton in 1971 showed a white haze over everything and grey clouds drifting out from the tall thin stacks. You can tell which way the wind was blowing that day. The cooling towers make large fat clouds, white and thick. The night-time photographs show deep grey clouds with orange halos around them, the factory illumination gleaming through. Sometimes the clouds were on fire.

Clouds of what? Ammonia made possible the feeding of the world. Synthesising ammonia from atmospheric nitrogen meant feeding streams of steam and air into gas generators that contained hot coke-made water gas and producer gas. The streams were purified and catalysed, and compressed and scrubbed and washed and converted. Massive amounts of natural gas provided hydrogen. Great amounts of carbon dioxide were produced as a by-product. Nature was captured and contorted and twisted into something akin to itself, but not quite itself, like a seeded cloud, made of humans, a relation of the wind and rain, but once or twice removed. How much energy was needed? How many jobs were made to undertake this action on air? How and for who? And all the other products and processes that made clouds. Who carried them out?

Who were the chemists and the engineers? Who were the chemical plumbers, the lorry and shovel drivers, the chargehands and plastics representatives, the electrical fitters, the laboratory technicians, the Works clerks, the messenger boys and girls, the typists and secretaries, the platers, the road-tanker loaders, the biscuit sorters, the gatesmen, the anhydrite miners? Who were the machine operators who prepared fibres, spun, twisted and wound threads? Who were the thread spinners and the cloud makers? Who were the thousands and thousands of people who fashioned the modern world from here?

In 1935, Bertolt Brecht wrote a poem titled 'Questions from a Worker Who Reads'. It transposed work, an everyday, continual activity, to the world of myth, ancient empires and the defeat at war of Philip II of Spain and the military victory of Frederick the Great. He asked, rhetorically, through the voice of the worker, who made the materials, who carried out the socially reproductive labour, which underscored and enabled the power of rulers? What were the names of the people that were lost to history, and yet, in truth, made that history happen?

> Who built the seven gates of Thebes?
> The books are filled with names of kings.
> Did the kings drag in the blocks of stone?
> And Babylon, destroyed so many times.
> Who built it up again and again? Of the houses
> In Lima, gold-glittering, which housed the builders?[1]

Brecht's worker asked where the stonemasons went the evening after finishing the Chinese wall and who erected the triumphal arches of Rome? The worker had many questions. Did the heroes of Ancient Greece win their battles alone? Who cooked for them? Who cried along with kings when fleets sunk under the sea?

> Each page a victory
> Who cooked the victory feast?
> Every ten years a great man,
> Who paid the expenses?

Brecht evoked the ancient world of Thebes and Rome, real places in which people worked and ate and fought and died. They were also home to the gods and the site of myths. Thebes was, according to legend, the birthplace of the mythical hero Hercules. Brecht's poem was written into his present, though. The worker who read and had questions about what has been read was developing a critical, communist frame of mind. From a knowledge of building, or cooking, of reproduction and social reproduction, questions arose about the handed-down, individualised heroic representations of the past. And the point was to make history class-consciously in the present.

Brecht's poem, which championed the questioning, self-educating worker, had little of the sensibility of George Orwell's rendition of the working-class attitude to education in *The Road to Wigan Pier*, from 1937.

> The time was when I used to lament over quite imaginary pictures of lads of fourteen dragged protesting from their lessons and set to work at dismal

jobs. It seemed to me dreadful that the doom of a 'job' should descend upon anyone at fourteen. Of course I know now that there is not one working-class boy in a thousand who does not pine for the day when he will leave school. He wants to be doing real work, not wasting his time on ridiculous rubbish like history and geography. To the working class, the notion of staying at school till you are nearly grown-up seems merely contemptible and unmanly. The idea of a great big boy of eighteen, who ought to be bringing a pound a week home to his parents, going to school in a ridiculous uniform and even being caned for not doing his lessons! Just fancy a working-class boy of eighteen allowing himself to be caned! He is a man when the other is still a baby.[2]

To leave school was to grow up, to be grown up, to earn money and avoid the teachers' beatings—even if there might be other beatings awaiting. Perhaps it was not in contradiction with Brecht's point about the questioning, learning worker, for he did not argue that schooling could provide Marxist wisdom, rather that only self-education or socialist party education was meaningful to the worker. But how was the worker that Orwell imagined made complicit with a rejection of education? To be a man was to be like a demigod, beating steel, hacking minerals in mine shafts, to build bridges and contribute to fiery life and glossy futures. Might each worker in Teesside be Hercules, the region's mythical patron. Might each at least exceed his seemingly proper place through the agency of new work and new opportunities?

When ICI arrived in the 1920s, it seemed that small improvements in living standards were possible. It was the promise of a future—new industry, new products, new jobs (See Fig. 1). Not everyone who dreamt beside its reactors, compressors and condensers, imagined a life inside the factory, but they could imagine a life of some kind. In 1937, social researchers from the Mass Observation project collected the thoughts of Middlesbrough High School boys, aged between 12 and 16, in 269 essays under the heading: 'When I leave school'.[3] These boys had stayed a little longer in education than many of their peers and wanted not ordinary jobs, but ones with a future and a good wage. The boys wanted careers, writing of securing, perhaps, 'a steady, well-paying job in an office' or of becoming 'an electrical engineer for a small firm such as Hoover or Goblin vacuum cleaners', or being employed in 'a job as a clerk or something that I can rise to a high position in, e.g. a lawyer or a customs official'. And one imagined working his way up in a big firm from 'office boy' 'head office' or perhaps becoming, like several others, 'a chemist in the ICI'. Many dreamt of security, of a pension and hoped for a good retirement. These were the better educated. That had to count for something. They knew about those who had lost out in the depression: 'We see these rough, idle and good for nothing men, hanging round the street corner with their hands in their pockets and we don't wish to be them when we leave school', wrote one. Life is precious—the sudden outbreak of war and then conscription, the coming of a strike, a deep depression all threaten the vision of a hopeful future. But others dreamt of lifting up and above the clouds, to see the world from the skies. One observed:

> To fly as a Royal Air Force pilot is my great wish for when I leave school. To defend my country in the latest inventions. To feel the thrill as you see Mother Earth rolling beneath you. To feel the breathless charm as you skim through the air at 400 miles per hour...If I was given my choice I would fly a single-seater racing monoplane, with a shining and streamlined body, and a roaring engine. And to have races with my fellow pilots, roaring over peaceful country villages, a contrast of old and new, climbing higher and higher and diving steeply, making it perform gymnastics at a touch of the rudder bar.

Another boy thought of lifting off from Middlesbrough into the skies above as a police constable in the flying squad or in the Royal Air Force, for he wished to 'fly above the clouds'. More prudently, though, he anticipated a life as 'a clerk in an office'.

Something had seemed to be shifting in the area. Official medical reports from the period noted a decline in mortality rates and incidence of infectious diseases from the end of the First World War.[4] Middlesbrough's Medical officer of Health, Charles V. Dingle, over-praised conditions in the city as 'exceptionally good' and attributed it to a decline in pollution matching the decline in industry, which had the benefit of improving health. Poverty, he noted, also meant a lack of money to spend on alcohol.[5] The closing years of war provided an opportunity for a new start, and so a detailed survey of the area was commissioned in 1944 and presented in October 1945.[6] As part of it, Ruth Glass and Griselda Rowntree carried out a social survey. The findings were bleak, the improvements proved to be illusory. Even the better areas were headed towards becoming as impoverished as the poor areas. One characteristic of the area was large number of elderly men, drawn there earlier for work, unmarried and unable to access health and care

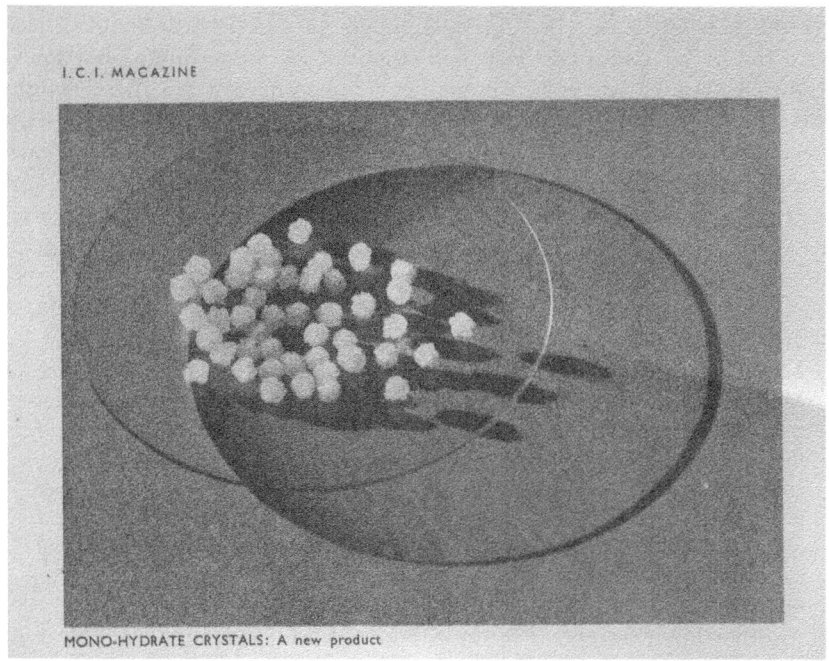

Fig. 1 A new product, Monohydrate Crystals, showcased in ICI Magazine, July 1933

facilities. Pollution from industry was noted as a cause of many ills. But the city needed industry and so that could not change. Perhaps a universal health service might ameliorate its effects.[7] Health centres connected to such a service, a National Health Service, did not come to the poor estates until the late 1960s. It was the past that had produced the bad present. As Glass's report put it:

> Not only are Middlesbrough's problems all intertwined but they also derive from one common cause: nineteenth century laissez-faire. Uncontrolled development has caused all the cumulative maladjustments; it has continually widened the social cleavages of the town. It cannot be allowed to persist.[8]

The city was split and splitting further. The seam was widening, like the widening of cuts through the rock, when the minerals are scooped out, used up. It broke apart, exhausted. The city was a problem. It needed to be planned. The state had to intervene. The science of the plan had to be applied. The Town and Country planning act of 1944, and its successor in 1947, bestowed on local authorities new powers to acquire lands for the world to come. The region needed a plan.

Planning and organisation was instigated at ICI too at this time (See Fig. 2). The techniques of labour management were applied to ICI's workforce in order to rationalise and mobilise the power of mathematics and labour science. An ICI booklet from February 1948, *The ICI Method for Job Appraisement for General Worker Jobs*, laid out, like the steps of a chemical experiment, how to make reliable formulas that bind wages to work.[9] The aim was a 'cohesive wage structure for standard method of job appraisement'. The booklet's instructions were extensive, for the process of assessment was laborious. It took half a day to assess a complicated job role. Modelled on procedures in place to assess 'general workers' jobs' at Billingham since 1937, it was adapted and extended to all ICI Divisions. The process demanded a bureaucracy of factory assessors, division assessors and inter-divisional assessors. The procedure was as follows: each job type was mapped against a series of main headings, consisting of Mental Characteristics, Physical Characteristics, Acquired Skills and Knowledge and Working Conditions. Mental requirements were particularly oriented towards the carrying of verbal messages, recording of simple figures and taking simple readings. The assessors asked of each role whether it required good memory, the ability to reason, a speediness of reaction and an even temperament—which was not to be confused with 'dullness'—any job which might require dependability under emergency conditions would rank more highly in this field. Each role would be rated against the need for co-operativeness, perseverance, a developed mechanical sense, which would not suit anyone who was clumsy, its demands for initiative and whether it required disparate attention—the ability to attend concurrently to a number of unrelated operations. Ability to visualise— to see the unseen mentally, to imagine the flow of fluids through pipes, to manage the translation of a drawing into the real world—was scored, alongside a sense of responsibility. Physical requirements related to the muscular strength, stamina, agility needed for the barrowing and moving of sand, sweeping of floors, walking across the plant sites, with exposure to outdoor conditions. In addition, the sensory needed to be assessed:

'Sensory Accuracy. Keenness of sight, hearing, touch, smell and taste'. Acquired skills and knowledge assumed that the datum line for education in such roles was 'below the level of learning of the highest class in a secondary modern school'. It considered the need and application in the role of training and the dexterity that comes from experience. Working conditions indicated the normal conditions of work in a 'good chemical factory: Conditions which after a week's experience of them pass unnoticed are considered to be within this definition'. The conditions were evaluated in relation to 'Physical Disagreeableness', including aspects such as vibration, cramped position, fumes, unpleasant smells, dust, dirt, heat, cold, wetness, inconvenient clothing, poor light. 'Mental Disagreeableness' arose from noise, working below ground, isolation, nervous tension and the risk of accidents and disease. The booklet had a special note on monotony:

> The characteristic described as monotony may apply where the cycle of operations recurs constantly at regular and very short intervals of time. Where the time of the cycle of operations is set by the speed of a machine not under the control of the operator the effect of the monotony may be increased. It does not apply where the cycle of operations is varied and repeats itself only after a considerable lapse of time; or to jobs which, whilst requiring an operator always to be in attendance, do not call for his undivided attention—such jobs might be described as dull, but not monotonous.[10]

Each job was scored against these factors, with the mental requirements and physical ones given a weighting of 0.4 and some extra emphasis given to 'Acquired Skills and Knowledge' (1.0) and 'Working Conditions' (0.7). The marks accrued were converted into money, or wages. The process was applied to adult men, as one category, and there was a variegated scale for women and juveniles.

In the same year, a system for 'The Recruitment, Training and Promotion of Technical Staff' was issued. Its approach to assessing its research chemists, design engineers, physicists and mathematicians and the like was more individualised than the 'test the job, not the man' approach for general workers. The booklet stated that the aim of the grading process was to 'obtain a picture of the value and progress of each member of the staff', as 'an individual in his profession', 'in his present job' and

THE HIGHPOINTS AND THE LOW ONES: IN, OVER ... 69

Fig. 2   The integrated company, from ICI Magazine, February 1938

'with regard to his long term potential'.[11] It assumed that, after assessment, it would be found that 20% were in the top A category, 60% in B and 20% in C. As many supervisors as possible were to appraise these individuals at regular points, grading them and plotting the scores on a graph, with the axes 'Years of service', which also logged age, and grading score, with annotations of milestones: 'Head of Test Section', 'Transferred to Research Dept.', 'Head of Design Group'.[12] A career life should become an upwards rising, occasionally spiky, line of unfurling development. This was the line of consistency in a world where things—and technical people—would only continue to improve. This work stemmed from Frank Ewart Smith, a former research engineer, who focused his attention in the post-war years to the science of labour management. He lectured in August 1948 on issues of industrial health, whose main problems he deemed to be psychological in nature. Aesthetic aspects of the factory could ameliorate the demands on the mind: lighting of a particular colour and without glare, restful hues on the walls. The senses could be cosseted even amidst the clatter of the factory.[13] If workers were conceived from the point of view of their bodily—and nervous—senses, so too might the modern factory be seen as an organism. In Ewart Smith's terminology, in September 1948, the 'central control rooms', which oversaw the increasingly automated functioning of the plant, were 'the heart and brain of an industrial process'.[14] These were a heart and a brain that were characterised, though, as tools or instruments. Like a heart ever beating, the central control rooms, the automatic processing allowed for continuous processing, under instrumental control, rather than batch or discontinuous processing. It signalled efficiency and consistency.

As the Wilton Plant expanded, apprentices joined an on-site training school, the third one to open in the company in twelve months. In the ICI magazine from August 1957, an article, 'Apprentice School', by Dorothy Thomas, of Metals Division Publicity Department, described the process of admission to ICI's new engineering training centres and what happened after acceptance. It wrestled with the differences between training and education.

> The words 'student apprenticeship' are perhaps misleading to the outsider. They do not, of course, mean that the beneficiaries sit with pencils or books in their hands throughout the whole of their training (most of them would argue that they rarely sit down at all!). The 'student' part is probably a gentle reminder that certain academic qualifications are expected.

> The recruits will, in general, have stayed at school rather longer than the craft apprentices, and they must have put the extra time to good use by gaining the G.C.E. at Ordinary level in at least four subjects, including mathematics, science and English.[15]

|The school at Wilton trained engineers for the future, in an environment described as 'a space-fiction panorama of great chemical plants where more than 10,000 people work'.[16] It was a 'nursery bed' of experiment and development.

> Every plant built here is dedicated either to producing something entirely new to Britain, like the 'Terylene' now in our shops or the titanium flying in our latest aircraft) or to developing for established products (nylon or polythene, for instance) a scale of manufacture never before attempted.[17]

In April 1958, Ewart Smith officiated at an opening ceremony for the Wilton training school. He supported the venture, noting that in his opening speech the necessity to develop training in Britain, for the Soviet Union was more advanced in this respect, with three times the ratio of teachers to population.[18] He addressed the apprentices with a lecture. It began with a metaphor about careers. A career should be seen as akin to a long trek in the mountains. It is always an ascent, but when it is believed that a peak has been reached, another will appear. The route is hard and not always straightforward. The enduring image illustrated by Ewart Smith was of a heroic path through a life that may be challenging but would always go upwards.[19] 'Life', he observed to the apprentices,

> personal life, family life and national life—does not stand still. It is either going up or it is going down, and when for a time it appears to be stationary, this merely represents the changeover from rise or fall or vice versa.[20]

He informed them about the principles that should guide them in all their technical—and managerial—work: Simplicity, Symmetry and Continuity. To this should be added Integrity and Wholeness. This was especially of relevance in an epoch in which work processes were fragmenting into many specialised parts. What bound it all together?

There was an ideal version of the factory and of the worker. The harsher reality was the ongoing concerns about danger and risk to life and limb, even in the modernising, clean, brightly lit and colourful factory.

An internal report to the Central Council on 1 May 1959 reported on the recent safety campaign.[21] Four fatal accidents occurred in the second half of 1958, bringing the year's total to five. Tables compared the number of accidents in different divisions. It reported on an Inter-Division Safety Competition to incentivise care at work. Types of accident included falls of persons and falls of materials. Causes ranged from human failing to system failings. Injuries were fractures and dislocations, contusions, strains, sprains, wounds and abrasions. Targets were set for accident frequency rates and tables provided a translation between deaths, injuries and lost man hours.

There were hazards in chemical work and so there was always a focus on health and safety. There was a duty of care in the factory and it expressed itself in curious ways at times. Women wearing nylon tights or stockings, likely made from ICI products, could be compensated if, as they passed through chemical processing areas and under pipe bridges to cross the plant, the acids and alkalis on the ground splashed onto their legs and burnt holes in the nylon.[22] But, as much as they were looked after, they were damaged too. Just one instance evokes the bright and colourful world of synthetics as it meshes with bodies. Workers at Wilton, processing Terylene, turned yellow from contact with an experimental process for dyeing, which also likely gave them cancer.[23]

Old and New Nature

> In feare and trembling they descend
> into threatened shock. Faire and
> softly, too far from the dry arbour.
> The chisel plough meets tough going,
> we spray off with paraquat 2 1/2 pints
> per acre. And the ^{51}Cr label shews
> them and us in your same little boat,
> pulling away from the vacant foreshore.
> J.H. Prynne, Poem 14, *High Pink on Chrome*, 1975.

Nature was remade by factory labour. It was and was not itself. There was a rift. Nature's offering was outbid through augmentation—synthesis. Life beyond the factory, consumer life, was to be remade. Synthetics could prolong the spring and bring it indoors for ever (See Fig. 3). But this meant that the earth was seeped with agrochemicals. Fertility was not natural abundance, but a result of a supplementation made necessary by soil exhaustion—which provided profitable business opportunities. Air was snatched in and puffed out into the atmosphere again. This constituted the industrial cycle. The nature around the factory changed. Labour was remade by the conditions and capacities of the nature that buffeted it and by the factory that contained it. Lives lived here adapted to the presence of synthetics. In March 1956, ICI Magazine reported on the birds around the factory at Billingham. Titled 'Birds of the Tees Estuary', an article, written by Charles W. Armstrong (Billingham Division), had a strap line remarking on how wildfowl and the production of chemicals seemed strange bedfellows, but 'it is a fact that within sight of the vast Billingham Works the lonely cries of the wild life of an estuary are still to be heard'.[24] Six bird types were illustrated by Raymond Sheppard: a Cormorant, a Curlew, a Herring Gull, a Dunlin, an Oystercatcher and a Mallard. Armstrong observed how, when the tide turned and crept across the mud, wispy clouds of knot and dunlins wheeled and twisted in amazing uniformity. Handsome oystercatchers and great black and ugly cormorants flighted and fished about. Describing the 'estuarine music', he noted that listening to fowl was a more fascinating pastime than watching them:

> Oystercatchers piped incessantly, and with them came the long, quavering whistle of whimbrel and piercing *tuck-tuck* of redshank. The melancholy cry of the curlew was symbolic of moor and estuary and all wild places; and harsh and frightening came the gabble of the fussy shelduck, a sound so peculiar that it has given rise to a legend in some fishing communities that it is the anguished cries of lost fishermen.
> Migrating widgeon whistled their peculiar soft whistle, which is heard into the Arctic Circle and beyond the Urals on almost every estuary of the Old World. More domestic sounding was the low quack of preening mallards, and the sounds were completed by the shrill pipings or trilling of the smaller waders.[25]

Towards the back of the journal, a darker note sounded. This was the company voice and it was more punishing. Under the heading, 'Battle of the Birds', a report began 'A new onslaught on the Billingham starlings began last month'.[26] The birds were coming inland, hanging around the factory. They were attracted, the article stated, by the warmth of the air above the pipe bridges and the smell of ammonia. Well over a million starlings, which, the report noted, 'make the night hideous with their piercing chatter'. And they left droppings that were inches thick

Fig. 3 Alkathene advertisement in the 1950s

all around. The report did not acknowledge the irony that dwindling supplies of those natural nitrogen-rich mineral deposits, made up of excrement, eggshells and the carcasses of dead seabirds—or guano—overly mined by indentured Chinese labourers and indigenous South American people—encouraged the whole quest for chemical substitutes back in the early days. In modern times, the birds were a nuisance. They had adapted too well to this new nature. They had to be shot. 'It is hoped', stated the magazine, 'that the shooting will demoralise the masses and drive them to roost outside the factory area'. Nature is to resettle out of place. It had to be displaced.

A more hopeful accord between bird life and factory was made in 1966: *Birds of Teesmouth*. Made by the RSPB, the 38-minute film examined the intermingling of birds and a thriving industrial port. It was a time of expansion, of steel production's glistening flying sparks and molten glowing yellow. Land was reclaimed from the mud, made solid from industrial tipping of waste. The land was built up of rubbish. And even here, on reclaimed ground around the factories, the film told the viewer, flowers may grow, such as banks of scentless mayweed, spotted orchids, colourful thistles, yellow wort. Slagheaps held a variety of wildlife— painted lady, common blue, meadow brown or tortoiseshell butterflies and more. Insect life was there. Turn a stone and you'll find a world of life beneath, it noted. The birds, stated the commentary by Anthony Clay, came in to feed on the mud in the river. We hear the birds' warbles, we hear how the birds' long down-curved bill pierced the mud deeply to extract worms. On Seal Sands, there were thousands of waders probing the mud. They were undisturbed by the passing shipping. Humans came to enjoy the sand or the promenade and the military parades at Seaton Carew. Rain appeared and the turbulent sea crashed against the pier. A pipeline from ICI fed into a reclamation pond and drew in birds, which were followed by James Monroe's camera across this *Kulturlandschaft*, wherever they wandered, including to 'the tips that are the ugly necessities of civilisation', where herring gulls fed. Even swans cruised through this humanised, industrialised environment. In winter, 'the slag heaps look like miniature Alps', and the reclamation pond iced over, cutting off access to food. Some birds were steeped in oil from the estuary, which stuck to them and caused them to die. If they rose to find more fruitful grounds, they might be tangled in the wires of the pylons, which were hard for them to discern, even though the electricity board had attempted to bring them into greater visibility. Snow buntings flew in to feed on sea wrack, amidst the surface coal swept up onto the beach and collected by local people to sell in small quantities. A black haze swelled around the factories, as the gulls and terns rested at the mouth of the river. Knot birds filled up the vista, like electrostatic flicker, ten thousand birds, grey and white, swirling as one. So many parts moving together, they flashed golden in the light, against a pale blue sky, 'sometimes eclipsing the sun as they rise and fall'. Industry was advancing from the background. New refineries were being built, steam gave way to diesel. Roads distributed industrial waste more efficiently across Teesmouth. Waste was now tipped into the estuary, killing two birds with one stone: waste was disposed of and made invisible and land was reclaimed. In the old days, the voiceover tells us, reclaimed wasteland was left alone, but now 'modern methods can fill in anything' and there is so much to get rid of. The mud flats, home of the birds, were threatened, for to 'modern industry it represents so much wasted space'. Two related questions are posed at the close of the film: how much space can we spare? How much do we want the birds to come here?

How much did we want life teeming uncontrolled in our environs? What price might be paid for more control? ICI introduced Paraquat under the brand name Gramoxone in 1962, having discovered its herbicidal properties at the agricultural research station at Jealott's Hill, Warfield in Berkshire in 1955. A powerful weed killer, just three drops of it were deadly to humans and other life forms. It explored the addition of an emetic, in order to induce vomiting in any one who ingested it. It considered adding a colourant to prevent it from looking like a soft drink or tea, and a stench to make it smell terrible. Such 'safety' measures were inadequate. Anything more was seen as too costly. It stayed on the market, despite concerns. Deliberate or not, it is estimated that the agrochemical caused tens of thousands of deaths, Furthermore, the developers knew it had the potential to accumulate in the brain, and could affect the central nervous system, perhaps leading to neurological disorders. Later records indicate that the company which later took over ICI's pesticides business, Syngenta, may have withheld from regulators the knowledge that its addition of a chemical, code named PP796, was not enough to counteract the fatal actions of the chemical. Furthermore, influencing strategies appeared to have been used to diffuse potential threats of banning.[27]

An augmented nature caused havoc elsewhere, an artificial fishiness in the air, engineered in the factories, and under discussion when the effects of factory production on the atmosphere around Billingham were the subject of debate in the House of Commons. On 29 July 1964, under the title 'Atmospheric Pollution, Stockton'. Bill Rodgers, the Labour MP for Stockton-on-Tees, raised 'a matter of great concern' to his constituents and one that he and his predecessor had been concerned with for a long time. It was, he observed, 'a specific kind of industrial pollution', and it appeared 'in Stockton, and which is of a very specific origin'.[28] He continued:

> At one time there was something which was known as the 'cat smell' on Tees-side. That appears to have been substantially dispersed, and what I am discussing is what is popularly known as the 'fish smell'. I cannot say that this smell emanates in such a way as to pass into my constituency alone and into no one else's. This I would regard as a rather single-minded exercise on the part of I.C.I. I think that at times it affects the constituency of my hon. Friend the Member for Middlesbrough, West (Dr. Bray) and also that of my hon. Friend the Member for Sedgefield (Mr. Slater), and I am sorry that it originates in his constituency, although I do not hold him responsible for it. There are times, also, when it is traceable as far away as Darlington.

Rodgers pointed out that Stockton suffered more than other areas from the smell, caused by amines from ammonia, because of the direction of the wind at certain times. Nature conspired further to plague the town: the sea fret or coastal fog, the result of a 'temperature inversion which makes what would be an unpleasant smell quickly dispersed something which persists and hangs over the town for a considerable time'. It was very peculiar and more unpleasant than other industrial smells and fumes. Rodgers quoted from some of the many letters he had received about the matter, producing a sensory archive of what it was like to live and breathe in the vicinity of ICI in the 1960s. One stated:

> We have just had what might have been two beautiful days completely ruined by haze and stinks, which have been particularly bad over Portrack, where I teach. I have had the misfortune to be developing a catarrhal cold and the discomforts of this have been greatly increased by the pollution of the air.

Another wrote:

> I would like to know how much longer the people in Stockton-on-Tees will have to suffer the horrible smell that comes over from I.C.I. I work in a nursery school and the only nice days of this year are spoiled for the children.

From a third: "My two children's health is suffering, as they have had terrible coughs for the past three or four months. The doctor cannot do anything about it as it is the fumes getting on their chests'. Another constituent stated that conditions forced him to take the decision to move out of the area.

The fishy smell irritated existing industrial diseases, such as chronic bronchitis. Rodgers drew evidence from a diary that had appeared in the newspaper, the *Northern Echo*. It recorded that, on 3 May 1962, the fish smell superseded the cat smell. ICI, at Billingham, had accepted responsibility, explaining that the smell stemmed from sporadic leakages from the methylamine plant. In July 1962, the smell hung over sixty square miles. A new plant was being prepared and it was said it would cut the leakages to a minimum—it did not. Again and again, there were insufferable seepages. The Alkali Inspectorate's Report noted leakages as a result of breakdowns and accidents, or units operating erratically and relief valves that needed to be opened—and the years rolled on. The chairman of Stockton Borough Council's Health Committee observed that, 'On every occasion when the atmosphere is heavy and the wind is in the north-east quarter we get it. That is the sum total of the matter: when the wind is in the north-east quarter and there is the right level of humidity, one gets this in Stockton'. Now and again, ICI accepted the work as the source of the smell. But along with the stomach-turning smell, there was complacency in the air. The Alkali Inspectorate, which was established in 1864 following the Alkali Act 1863, operated under the principle of 'prudent tolerance', which balances the needs of industry with the protection of the public. Rodgers pushed responsibility back to the government:

> I do not hold I.C.I. responsible for this nuisance. It is its job to manufacture and sell, but it is the job of the Ministry to control and not expect that its simple moral authority will require a large private firm to do what might be in the best public interest.

Rodgers reminded Parliament, in summary, that Stockton-on-Tees helped to pioneer the Industrial Revolution a hundred years before, but had now become the victims of that expansion. It was the case that the factories mingled with the mists and trapped emissions and the emissions slowed down the dispersal of the mists—the air was thick and smelled of fish. The Parliamentary Secretary, Mr Corfield, pushed the problem back to those who lived there—their domestic production of smoke was equally to blame. He assured the House that, in time, the engineering problem would be resolved, but could give no time scales or guarantees. 'I am not a chemist, and not a very good engineer', he stated, but even so, he understood that the situation was not straightforward. Solutions were hard to find. Work had to go on. There were clouds to be made. Threads to be spun. Fertile dust to be compressed. Orders to fulfil. History to be made. Nature to be remade.

A Plan

One hundred years after Gladstone had named the town of Middlesbrough, which he saw as an infant of industry, Hercules, the ground and the water and the air off which that child had thrived were polluted and the future uncertain. A plan was initiated. The Teesside plan was developed in the 1960s to make a city fit for the new millennium to come. Architect and town planner Franklin Medhurst, a man who was called on to develop the plan for the region, wrote of his first view of Teesside, in 1965, enveloped in a haze of pollution, sun diminishing:

> The sharp line of the Cleveland Hills marked the northern fringe of the North York Moors as we approached the long, hardly perceptible descent into the Tees valley. At the same time we were conscious of a slight hazing of the crystal sunlight, slowly thickening until some fifteen miles from the valley towns we were driving in mist with the sun's orb rapidly dimming and an obvious chill coming through the car's ventilation system.[29]

By the time the car had reached Stockton-on-Tees—with his family, come to move into the area—there was a 'fetid fog'. Its density reminded him of his first encounter with the area—from the skies, on an RAF U-boat hunting mission in the North Sea, just after the war had ended. His aircraft was surrounded by grey, then tawny-yellow orangey-brown and black clouds, 'clouds of hell', the clouds over Teesside.

> We were over a grimy, gloomy metropolis of 'burning sands and eternal storms', smoke and fog-bound, pierced by fountains of flames from flare-offs, immersed in acrid fumes which made our nostrils twitch as they filtered through gaps in the fuselage.[30]

From the air, it seemed to him like a realm of the damned, as imagined by Dante, but it was a city of flaming furnaces, which had forged the weapons of war, the ships and more structures in steel and iron. In amongst this industry, almost inconceivably, he thought, were people, the workers, eaters, sleepers, shoppers, players and, more than anything, the cleaners, who dealt with the daily smutting of washing on the line, the dusty grit that blew in, the filthy clothes. And what about that pollution that could barely be seen?[31]

In 1965, the airman was a car driver and he approached the fog from the ground. He came to find ways to lift the murk, or eliminate it, to give the towns a centre and something approaching a civilised future. The old Ironmasters district between Stockton and Middlesbrough, which had been aflame with production, was derelict now. The roads were potholed, the metal gates rusting. Hercules was a fateful model to choose for Middlesbrough. While he was stronger than any mortal, his strength meant that he willingly endured hardships and he suffered many setbacks. Hercules died an untimely death, tricked by a centaur who managed to get him clothed in a poisoned cloak. What poisons had cloaked Teesside over the years? What stories and myths of the dignity of labour? How to honour the work and dedication and ingenuity of those heroes of industry, those on the shop floor, in the plant, under the ground, without the hardships, the setbacks, the suffering and the unfairness. And while the workers may have vacated the area around the Ironmasters district, as those ironworks and engineering works closed down, they still made their journeys to the edges, to the town's penumbra, its cloudy suburbs, where new clouds were made—by the deep-water Teesport or the ICI plant at Wilton. These places were somewheres that seemed a galaxy away from the poorly towns, corporate, globalised suburban spaces, places where routes were arc-lit and where illegible industrial structures were floodlit, and where, yet again, raw materials were manufactured and exported elsewhere, for further processing, so wealth was made and went somewhere, if not here. This was an area made by modernisation, commandeered under an ideology that believed in progress, through new industries and new methods. and long advocated by regional industrial development boards.³²

Modernisation had formed an economy built on chemicals and steel, transformed by state-subsidised investment and hi-tech, in a built environment that was planned and where public expenditure was directed towards the provisions necessary to modern existence: roads for cars, houses for workers, schools, training centres, health facilities and shopping centres. Stockton High Street was modernised in 1960. Victorian shops gave way to a shopping centre. Modernisation was delivered by the modern factory products. An ICI promotional brochure, *Signposts 6*, from Plastics Division in the early 1970s showcased the uses of Perspex in shopping centre signage, it cited Clifford Barnett from his talk 'Less is more', on large area glazing with acrylics: 'I believe that we should now aim to provide less architecture in our building and try to create more environment'. Environment, the brochure writer elaborated, is 'where the action is'.³³ Environment was a social nucleus with a promise of something to come, a somewhere that draws people in. Bricks and mortar, solidities, had given way to accessible walk-through settings, well connected for cars, bus stations, walkways, and it was produced in and through social amenities, commercial services, places where buildings were arranged to serve, not dominate the environment. Still these new spaces should not be without individuality. Here Perspex, or the rigid PVC sheet Darvin, played their role. They supplemented the 'instant buildings' of 'instant environments', modular and apparently practical on the drawing board, but repetitive and impersonal translated into concrete and cement. Illuminated plastic signs 'bring facades, precincts, car parks and shopping malls to life'.

In this lively plastic modernity, wealth was made, but it did not settle much in the vicinity, unlike the pollution, which laded and seeped into the ground and hung around. And dirt attracts dirt. Waste incinerators that would be out of place elsewhere came to join the other dirty industries. But there were countervailing tendencies with the switch from coal as fuel and feedstock to Naphtha, a cheap low-grade petrol, in 1963. And yet pollution still lingered in the air.

In 1966, Stockton Borough submitted a planning application—new homes were to be built at Haverton Hill in Billingham, and it was found that the sulphur levels there were very high, so high indeed, according to the County Analyst for Durham County, that 'when it rains at Haverton

Hill it rains dilute sulphuric acid'.[34] It was popularly said that even the birds coughed there.

The airman, Franklin Medhurst, devised a plan, the Teesplan. Teesside was unique on earth, it seemed to him. The towns were dying, mouldering under the fug of pollution, the waters polluted, the ground contaminated and de-industrialisation left hulks of factories and warehouses, amidst what had been attractive Georgian and Gothic Victorian and Edwardian streets and buildings, built along a meandering river. Productive activity had moved to the edges, to the new port, to the new factories. There was production. There was work, more and more of it, but it had not urbanised the area in the conventional manner. Rather it had left the older coherent towns to rot. In the *Middlesbrough Evening Gazette*, in March 1977, Medhurst, wrote the following:

> In no other region of Britain or elsewhere can you find a renaissance in industrial growth without a renaissance of the urban structure. Yet this was the situation in the six towns. Surging busy industrial centres sat cheek by jowl with the torpid, weary indolence of towns.[35]

Stockton-on-Tees had been a handsome town that grew up with the industries that emerged there. It had a fine town hall and a music hall, a theatre, an impressive hotel, remarkable inns. These were historic meeting places, mainstays of community life. The airman had a plan that would hold on to what was worth holding on to and to bring back a sense of coherence. In 1969, he co-authored a book, *Urban Decay*, which proposed ways of remediating the deteriorating fabric and functions of cities and towns, rapidly built nineteenth-century stock, for the most part.[36] His own interest in this question related to the house he bought in Stockton-on-Tees in 1965, which had been the home, for eighteen months, of George Orwell, Eileen and their adopted son. Here, claimed Medhurst, Orwell began to write *1984*, then called *The Last Man in Europe*. He suggested to Stockton Council that they might list this Georgian-Victorian hybrid. The response: 'We don't want any more historic buildings. We have enough as it is'.[37]

What to do with the old and the new? Medhurst suggested that the towns of the Tees Valley were to work in unity, to thread the tangle of pearls strung along a coiling waterway into a conurbatory chain. There would be order, according to need and logic. Stockton was to become the pedestrianised historic centre, for shopping and entertainment. Middlesbrough was to be the regional centre. Redcar was to become a summer leisure resort. Much of the plan—including the proposal for a unitary authority as city, with a distinct focus in each of the towns and its recommendations around pollution alleviation and substandard housing—was junked, along with the planner, who was required to leave the area under a cloud. A reorientation towards demolition and identikit rebuilding, the imposition of a dual carriageway, cutting off the river, and the requirements of a modernising industrial economy took place under cover of another fog, a still hazy one comprised somehow of corrupt planners and councillors, dealing behind firmly closed doors, government interference and cost-cutting.

Perhaps the plan assumed that it had as much right to determine what life was and could be in the area as other forces that had a stake there. The chemical companies made the atmosphere, provided the means to sustain the self, seeped into all areas of life. They were in themselves an atmosphere. Journalist Martin Walker wrote an article in *The Guardian* on 23 March 1983, titled 'The Region Betrayed by False Gods', on the 'new feudalism' of the modern multinational corporation, with its cradle-to-grave paternalism, its feudal livery, its own code of loyalties and rewards:

> If you work for ICI along the River Tees you can buy your wines on the ICI label, fill up your car at the ICI garage, buy your clothes and suitcases and bedding at cut rates at the ICI shop. If there is no work for you, you can draw your pay and spend your time watching videos on ICI's Mickey Mouse shift. Your local newspaper will be ICI's *Billingham Post*, and for senior employees, school fees can be paid by ICI scholarships.[38]

The un-towns were company towns. The company was a permanence that pulled in labour from all around, until it did no more. The company was the towns and made the towns and their social structures. Who could challenge that? Khadim Hussain, worked at ICI Wilton, in Research and Development on process optimization from September 1977 to September 1991. He recounted how he got the 9–5 job as a laboratory technician.

> Long story, I came to England at the age 10, not speaking a word of English, my secondary schooling was in Bradford. Due to problems with

the English language I was pushed towards science subjects. I took my 'O' in 1975 and for the first time a teacher was appointed (rather press ganged). He offered a 15 minute appointment to anyone interested.

When I saw him, he looked at the subjects I was doing, English, Maths, History, Physics, Human Biology, and suggested I apply for job in the laboratories of the local Textile mills and added 'What a pity. If you lived in Teesside, thousands of Jobs in ICI'.

I moved to Middlesbrough in summer 1975, saw a careers officer and obtained the address for ICI and applied for a job. The reply was sorry no job at this moment—we've kept the application on file. If you do not hear from us, after 3 months, apply again. I had the date to reply marked on the calendar. Finally in July 1977 I was called in for an interview and was successful.[39]

Hussain's job was in the laboratory, not on the factory floor. The divisions were palpable, in his recall.

We worked in small groups of three or four maximum, with contacts with the group when necessary. The people working there were mainly from Marton, Nunthorpe, Guisborough, New Marske, Saltburn, the areas I regarded as middle class. As far I can remember there was no one from Dormerstown, South Bank, Grangetown, Lazenby or central Middlesbrough.[40]

And even as technical staff still he could not escape the effects of class society:

I had very little or no contact with anyone out of work which could be due to:

Because I did not drink or

At that age I lacked social skills.

I was from a working class.[41]

Sociology and Chemistry

In 1977, industrial sociologists Huw Beynon and Theo Nichols wrote *Living with Capitalism: Class Relations and the Modern Factory*. The book outlined working life in a fictional factory called ChemCo, described as a Riverside plant. Here pearls of nitrate fertiliser were made. It could have been ICI at Teesside. If names and locations were blurred and hidden, the interviews were real and were recorded between 1970 and 1973. This ChemCo plant was run by the then newest process technology of the chemical industry, a system whereby production takes place ceaselessly within enclosed vessels, which are monitored and controlled by the manipulation of levers. In the process production at ChemCo, the monitoring and adjustment function was presented as an enriched labouring environment, cleaner, more skilled, less physical, and, most importantly, needing fewer workers and vastly augmented productivity. In the real world, in the 1960s, ICI had built new plants at Billingham and Wilton according to such modern methods. These formed at this point the biggest petrochemicals and general chemicals complex outside of the US. The polythene plants were the third largest in the world and the biggest in Europe. Here was the massive processing of materials in sealed machines. Technology changed the social relations in the factories. The new processes with 'large single stream units' seemed to demand more centralised control, and there was more at stake should something go wrong. Before there had been smaller plants, which evaded tight control but also did not produce disasters if something went wrong.[42] The automated processes led over time to the large increase in outputs and a diminishing of the workforce. To be a worker here was to repeatedly observe gauges and dials, note down temperature, pressures and rates of flow and record these on log data sheets. The interviews in the book on ChemCo cloud the sunny vision of process technology. They gave details of tough physical labour, the dragging around of raw materials and the sacks of fertiliser, the finished product. They bore witness too to the loneliness and stress of the control room workers, left alone with their thoughts and the noise of the plant, at all hours of the day and night. The shift system was disruptive to life patterns—there were shifts of eight hours, from 06:00–14:00, 14:00–22:00 and 22:00–06:00 and these rotated throughout the week. It might be that a worker did the 06:00–14:00 shift on Monday and Tuesday and, then, on Wednesday and Thursday, it would be the 14:00–22:00 shift, before moving once again to a 22:00–06:00 shift for Friday, Saturday and Sunday. Monday and Tuesday afterwards were days off.[43] Perhaps the process technology system, and the way it had to be managed had something more in common with a monstrous foreboding of it, as seen in Fritz Lang's 1927 film *Metropolis*, when an exhausted and panicked worker overstretched himself again and again, moving the hands on a clock-like machine, to follow the impulses of rapidly flashing light bulbs.

Beynon and Nichols wrote about the dust that was so thick the workers could no longer be seen and they mentioned that at Teesside ICI there was just such a store where, as *The Guardian* reported, after a court case in 1973, the dust was two feet deep and there were 'fertiliser dust stalactites hanging from the roof'.[44] The punishment for this breach of the Factories Act 1961 was a risible £50 fine. On 21 January 1969, a blast ripped through the air of Teesside. It stemmed from the Polythene No. 2 plant at ICI Wilton. A cloud of cyclohexane gas leaked out and suffused a diesel high-lift truck. Though the driver acted swiftly to switch off the vehicle, the gas was sucked into the engine and exploded. Workers were injured and some died. To be in the newest factories, with process technology, was not always a clean and controlled process. Ron Angel's 'Chemical Workers Song', from 1964, stressed living and breathing or dying and choking, amongst all these clouds of gas and dust and pools of toxicity. The lyrics spoke of how every day in the place took a worker two days closer to death.

> A process man am I, I'm telling you no lie
> I work and breathe among the fumes that trail across the sky
> There's thunder all around me, poison in the air
> A lousy smell that smacks of hell, dust all in my hair.[45]

Amongst the spinners the smoke is oily. Shovelling gypsum makes you choke. The caustic burn of cyanide makes you sick. The money is decent, especially with overtime, and the bonuses are good, but it speeds up the passage of life and delivers you to death's door before your time.

The managers in their interviews with Beynon and Nichols, however, gave voice to their sense of logic and rationality in overseeing the factory's operations. They seemed blind to their existence in an irrational market. Modes of management were changing in the period observed. The New Working Arrangement, a nationally negotiated deal, was introduced in the late 1960s, and operated effectively to curtail the power of the trades unions, through their incorporation into the structure. The deal disallowed local pay negotiations at the plant level, instead introducing a nationally negotiated pay system, based on a classification within seven grades and calibrated, by managers, against a set of qualities that were awarded points, which were used to determine recompense. The qualities were Memory, Visualisation, Original Thinking, Disparate Attention, Even Temperament, Co-operativeness, Leadership, Agility and Sensory Accuracy.[46] At a local level, it had to appear that the union was strongly involved in the world of the factory. At ChemCo, a check-off system for union subscriptions was brought in, dues taken directly from wages by the employers. There was the introduction of a closed shop, which meant all workers had to be bound into the union, which, in turn, meant that there was 100% union membership, but its existence was for the most part on paper and it made it less likely that there would be unofficial or wildcat actions. ChemCo's managers took an interest in who was elected as union shop stewards—with foremen encouraging certain candidates to stand, especially those who had never had previous involvement in trades unionism before coming to work at the site.[47]

The new working arrangements, in reality titled The I.C.I. Weekly Staff Agreement, included an increased productivity deal.[48] ICI was angling to become the largest textile manufacturer in the world and was buying up other companies. Sir Peter Allen, the chairman, had been on a fact-finding mission to the US, where DuPont was the major competitor to ICI. He claimed that the US firms were able to use two-thirds less labour to achieve the same ends. The workforce of 55,000 people would agree to removing 'restrictive work practices', such as job demarcations, overtime and merit payments. Workers would aim for a 15% productivity increase, and, in return, receive wage increases of up to 22%, weekly instead of hourly pay, the abolition of the time clock, ability to have a tea break at will and common rooms on the shop floors.[49] Management claimed that, without the deal, there would have to be redundancies.

Resistance to the new accord was partial—and in any case, the organised workers of the chemical industries had long tended to work in partnership with management. Sir Alfred Mond, first chairman at ICI, had argued that, 'the best answer to socialism is to make every man a capitalist'.[50] What he meant by this was to bind the workers by whatever means—ideological, structural—into the success of the system that exploited them, such that they would consider it their duty to make its frictionless functioning in their interest. To this end, Mond had introduced 'perks', which were seen as 'voluntary acts' on the part of the employer, not a result of trades union negotiations: share-ownership schemes, holidays with pay, company housing, a company magazine and recreation clubs.[51] This was how you forged company men. Also, in its earliest days, from 1929, ICI formed Works Councils for labour-management cooperation and trades union membership at ICI fell until the 1950s, in contrast with other sectors. By the postwar period, these workplaces had none of the militancy of coal miners or engineering workers, against whom the new working directives were primarily targeted, in the coming period of increasing rebellion. Strikes were always uncommon at ICI. Works councils and joint consultative committees negotiated deals without militancy occurring. In 1975, there was a craft union strike at Billingham, which lasted three weeks, but it was resolved informally.[52]

The teeth of the union were pulled at fictional ChemCo and real world ICI. But they were about to fall out anyway. Work was changing—administrative and service sector expansion, call centres, tax and social security processing and less heavy industry. There were fewer employees in the heavy industry that did exist. There was more precarity. Few jobs for life—this was a time of flexibilisation, self-making and remaking. No living within the paternalistic cradle of the super-employer who made your bowls, curtains, cutlery, clothes, pills, paints, washing powders, the fertiliser that grew your food—and provided a social life and structure, with clubs and holidays and beanos and training. No more long service milestones.[53]

The airman in the sky, once he had emerged from the bottom of the weather cloud that became a fog of pollution, saw the flaming factories, but what he thought about were the workers who kept them alight, and smashed and extruded and spun and drove and smelted and gathered and sorted and dug the matter and materials, to make products. He also thought about those who played and learned and cleaned amongst the flames and the ash and the dust, which is to say he imagined the penumbra or cloud around the factories' labour. The cloud in the factories' midst covered those who did not work at the factory, housewives, children, pensioners. They did not work in the factories, but they smelt what came from them. They breathed it in. It was their atmosphere too. How did they live amongst it? What was it like to live here in transformed wind and water? How did the air smell today? Was the wind acrid? Was it better than yesterday? What sounds escape from under the ground? How far could we see down this street? Could you hear gushing water around the plant? Did the chemicals in it modulate the sounds or did it sound like the purest stream of water swishing through? Are you coughing now? Are you smothered? (See Fig. 4).

Notes

1. Brecht, *Werke*, BFA 18, Suhrkamp, Frankfurt/Main, 1988–2000, p. 409. Author's own translation.
2. George Orwell, *The Road to Wigan Pier*, Penguin, Harmondsworth, 2021, p. 80.
3. These are discussed and excerpted in Hester Barron, Claire Langhamer, 'Children, Class and the Search for Security: Writing the Future in 1930s

Fig. 4 Bottle of Trichloroethylene, England, 1940–1960 (Science Museum, London)

Britain', *Twentieth Century British History*, Vol. 28, No. 3 (September 2017): 367–89.
4. Kate Nicholas, *The Social Effects of Unemployment in Teesside: 1919–1939*, Manchester University Press, Manchester, 1986, p. 80.
5. Kate Nicholas, *The Social Effects of Unemployment in Teesside: 1919–1939*, Manchester University Press, Manchester, 1986, p. 81.
6. Ruth Glass, *The Social Background to a Plan*, Routledge and Kegan Paul Ltd, London, 1948.
7. Ruth Glass, *The Social Background to a Plan*, Routledge and Kegan Paul Ltd, London, 1948, pp.81–2.
8. Ruth Glass, *The Social Background to a Plan*, Routledge and Kegan Paul Ltd, London, 1948, p. 187.
9. 'The ICI Method of Job Appraisement for General Worker Jobs', Feb 1948, Copy held in Institute of Mechanical Enginner's Frank Ewart Smith archive: FES/3/1/1.
10. 'The ICI Method of Job Appraisement for General Worker Jobs', Feb 1948. Copy held in Institute of Mechanical Engineers' Frank Ewart Smith archive: FES/3/1/1, p. 10.
11. 'The Recruitment, Training and Promotion of Technical Staff', 2 Mar 1948. Copy held in Institute of Mechanical Engineers' Frank Ewart Smith archive: FES/3/1/3.
12. 'Staff grading chart', c.1948: Copy held in Institute of Mechanical Engineers' Frank Ewart Smith archive: FES/3/1/2.
13. 'The Contribution of Engineering to Industrial Health', 17 Aug 1948. Copy held in Institute of Mechanical Engineers' Frank Ewart Smith archive: FES/3/1/4.
14. 'The Contribution of Instruments to Industrial Progress', 14 Sep 1948. Copy held in Institute of Mechanical Engineers' Frank Ewart Smith archive: FES/3/1/5.
15. ICI Magazine, August 1957, p. 258.
16. ICI Magazine, August 1957, p. 254.
17. ICI Magazine, August 1957, p. 258.
18. 'Wilton Works Training Centre', 30 Apr 1958. Copy held in Institute of Mechanical Engineers' Frank Ewart Smith archive: FES/3/1/8.
19. 'Address to apprentices', 30 Apr 1958. Copy held in Institute of Mechanical Engineers' Frank Ewart Smith archive: FES/3/1/7.
20. 'Address to apprentices', 30 Apr 1958. Copy held in Institute of Mechanical Engineers' Frank Ewart Smith archive: FES/3/1/7, pp. 8–9.
21. 'Safety Department Report to Central Council', 1958–1 May 1959. Copy held in Institute of Mechanical Engineers' Frank Ewart Smith archive: FES/3/1/6.
22. As told in an interview by Nicola to Jon Warren, quoted in *Industrial Teesside, Lives and Legacies*, Palgrave Macmillan, Basingstoke, 2018, p. 157.

23. As told in an interview by Anne to Jon Warren, quoted in *Industrial Teesside, Lives and Legacies*, Palgrave Macmillan, Basingstoke, 2018, p. 159.
24. ICI Magazine, March 1956, p. 74.
25. ICI Magazine, March 1956, p. 74.
26. ICI Magazine, March 1956, p. 89.
27. https://unearthed.greenpeace.org/2021/03/24/paraquat-papers-syngenta-toxic-pesticide-gramoxone/
28. All quotes relating to Bill Rodgers are from his speech 'Atmospheric Pollution, Stockton', *Hansard*, Vol. 699 (29 July 1964).
29. Franklin Medhurst, *A Quiet Catastrophe: The Teesside Job*, Citizens Papers, 2010, p. 8.
30. Franklin Medhurst, *A Quiet Catastrophe: The Teesside Job*, Citizens Papers, 2010, p. 10.
31. Franklin Medhurst, *A Quiet Catastrophe: The Teesside Job*, Citizens Papers, 2010, pp. 10-11.
32. Jon Warren, *Industrial Teesside, Lives and Legacies*, Palgrave Macmillan, Basingstoke, 2018, p. 69.
33. *Signposts 6*, ICI Plastics Brochure. Author's own collection.
34. Franklin Medhurst, *A Quiet Catastrophe: The Teesside Job*, Citizens Papers, 2010, p. 52.
35. Franklin Medhurst, *A Quiet Catastrophe: The Teesside Job*, Citizens Papers, 2010, p. 112.
36. Franklin Medhurst and J. Parry Lewis. With a Chapter by Elizabeth Gittus, *Urban Decay: an Analysis and a Policy*, Macmillan, London, 1969.
37. Franklin Medhurst, *A Quiet Catastrophe: The Teesside Job*, Citizens Papers, 2010, pp. 45–6.
38. Cited in Franklin Medhurst, *A Quiet Catastrophe: The Teesside Job*, Citizens Papers, 2010, p. 130.
39. Personal correspondence with author, 18/05/2020.
40. Personal correspondence with author, 18/05/2020.
41. Personal correspondence with author, 18/05/2020.
42. Andrew Pettitgrew, *The Awakening Giant: Continuity and Change in Imperial Chemical Industries*, Routledge, London, 1985, p. 199.
43. David Walker, *Occupational Health and Safety in the British Chemical Industry, 1914–1974*, Doctoral Thesis, Strathclyde, 2007, p. 27.
44. Huw Beynon and Theo Nichols, *Living with Capitalism: Class Relations and the Modern Factory*, Routledge and Kegan Paul, London, 1977, p.12.
45. Ron Angel recorded a version for The Tees-side Fettlers, *Ring of Iron*, Traditional Sound Recordings TSR 016 (LP, UK, 1974). The LP notes state: 'Written by Ron, about 1964, the background of this lament is not the harrowing and dramatic accidents which can happen in a chemical works, but the process worker's awareness of the long-term efforts of a

working environment polluted by excessive noise, dust, noxious fumes etc.'
46. Discussed by John Charlton, *Socialist Worker*, no. 155, 22 January 1970, n.p.
47. Huw Beynon and Theo Nichols, *Living with Capitalism: Class Relations and the Modern Factory*, Routledge and Kegan Paul, London, 1977, p. 115.
48. Pramod Verma, 'Collective Bargaining in the British Chemical Industry: A Case Study', *Indian Journal of Industrial Relations*, Vol. 8, No. 4 (1972): 513–26.
49. Summary from *Business Week*, 9 September 1972, cited in Edward Maynard Glaser, *Improving the Quality of Worklife…and in the Process, Improving Productivity: A Summary of Concepts, Procedures and Problems, with Case Histories*, Human Interaction Research Institute, United States, Department of Labor, Manpower Administration, 1975, pp. 76–7.
50. W. J. Reader, *Imperial Chemical Industries, A History, Volume II, The First Quarter-Century 1926–1952*, Oxford University Press, 1975, p. 60.
51. Colin Gill, Ralph Morris and John Eaton, *Industrial Relations in the Chemical Industry*, Saxon House, Farnsborough, 1978, p. 87.
52. Andrew Pettitgrew, *The Awakening Giant: Continuity and Change in Imperial Chemical Industries*, Routledge, London, 1985, p. 200.
53. See Eloisa Betti, 'Historicizing Precarious Work: Forty Years of Research in the Social Sciences and Humanities', *International Review of Social History*, Vol. 63, No. 2 (2018): 273–319.

ICI and the Senses

Abstract This chapter charts the stellar rise and then subsequent break up of ICI over the 1980s and 1990s. In those final years, in the context of globalisation, there was a shift in emphasis from bulk commodity chemicals to higher growth, higher margin chemicals. A postmodern-inspired consciousness of branding and consumer markets meant that a rhetoric developed that focused on sensory gratification. This was a, presumably unintentional, echo of the smells and feelies in the dystopian novel *Brave New World* (1932), written by an early company employee, Aldous Huxley. What affects the chemical made by the company had on the body is followed up in this chapter in a different way, through an exploration of the fatal industrial diseases that plagued those who had worked in the factories, many of which came to light only at the end of the company's life. The chapter closes with a reflection on memory of ICI in the Tees valley region and how it is communicated through the channels of social media and meshes with a history of conspiracy and secrecy that has shadowed the works in the region, and continues into the days of the new freeport to be built in the region. The chapter includes reflections on what remains of all this, including what toxicities—and asks if even the clouds—and the stars—have been transformed by the factory.

Keywords Artificial flavourings · Brexit · Divestments · Freeport · Globalisation · Mesothelioma · Quorn · Sensism · Teesport · Teesworks

© The Author(s), under exclusive license to Springer Nature Switzerland AG 2023
E. Leslie, *The Rise and Fall of Imperial Chemical Industries*,
https://doi.org/10.1007/978-3-031-37432-6_3

End of Days for ICI

ICI was a powerful company from its beginnings in the 1920s and all through the years of war. It got on to a strong footing in the period following the Second World War. As decolonisation occurred and, across the world, trade connections and markets recomposed, it held its own against major competitors in Germany and elsewhere. ICI's fortunes mirrored that of the wider UK manufacturing base, of which it was a substantial part. It was a bellwether company: if the going was tough at ICI, it would be hard going across manufacturing in the UK and this would be reflected in its share values. The company's fortunes—literal tangible monetary fortunes—peaked in the 1980s. John Harvey-Jones was leading the company in 1984 when it made an unprecedented £1 billion of pre-tax profits. It was in this decade that the company tagline changed, from 'Our greatest asset is our people' to 'Our greatest asset is our customers'.[1]

ICI had been a company that, like other vast company conglomerates of the first half of the twentieth century, such as IG Farben in Germany, had been built up on the basis of vertical integration of all its parts. It was a company and all were combined within it, from worker to Divisions (see Fig. 1). From its large headquarters in London and an intricate arrangement of control groups and committees, ICI had managed an array of products and processes through its product divisions—mostly heavy chemicals—and sold and manufactured these in the UK and the Commonwealth. Substantial investment in research and development meant that there were constantly new products and new processes coming on stream. Business lines that had become outmoded were replaced by new ones from across the company. Plants grew, and in the process became sprawling complexes, with new facilities bolted on. Product streams proliferated, as Divisions picked up on the inventions developed in another area and developed uses for them. ICI pushed into European markets in the 1960s and US ones in the 1970s. In this decade, things began to change. Tariffs were reduced, and so new players entered the scene. There were fewer innovations and rates of growth slowed down. The large company, with its concentration of power in its headquarters and its complex array of divisions, found it hard to react swiftly to broader events or to change direction quickly, in response to developing circumstances. But, eventually, change was forced on it by an all-encompassing circumstance, when economic recession beset the UK economy in 1980–1981. Harvey-Jones came in, or rose up spectacularly, having been a trainee work study manager on Teesside in 1957 and on the board of directors from 1973. He stripped back the management structure, breaking up its centralising aspects, to allow fast reaction times at its edges: speed rather than direction is said to have been his motto. This former seaman and naval intelligence officer claimed that the winds of opportunity could always be mobilised to veer and tack as needed. ICI undertook a plastics swap with British Petroleum in 1982, exchanging polyethylene for polyvinyl chloride (PVC), and so it moved away from a historic core product line, one that had been of its own invention. Plastics, fibres and industrial chemicals were combined in a subsidiary called ICI Chemicals and Polymers and remodelled through joint ventures, divestments and product swaps with other companies. Years of recomposition began.

Under Harvey-Jones's leadership from 1982, ICI cut back operations, shrank the corporate headquarters, laid off tens of thousands of workers, reducing the workforce by a third. It looked for ways to utilise the assets. The salt mines underneath the Billingham site—closed in 1978—were earmarked as a repository for nuclear waste in 1985. A material with long-lasting agential capacity was to be sealed into the scene. Protests scuppered the plans. Efforts were made to rearrange production in the face of competition. For example, fertiliser was no longer so profitable. In 1981, Norsk Hydro bought the UK fertiliser business Fisons and appeared to be a more competitive outfit. The price structure was threatened by cheaper fertiliser from Europe, which was poised to take over the UK market and elsewhere. Urea with much higher nitrogen content was exported from the USSR. ICI bought up other British fertiliser manufacturers and considered moving its manufacture of methanol to places where natural gas is relatively cheap, such as the United Arab Emirates. But demand for fertiliser was anyway falling, as increased food production in Europe led to grain and beef 'mountains'.

The image given off by the company was one of enduring stability. In 1988, corporate identity consultancy Wolff Olins redesigned the ICI roundel: the two white lines representing the sea were no longer wavelike, but rather almost level. Seas should seem calmer. Two years earlier, the same brand agency had redesigned the Labour Party logo, dropping the word party and replacing the red flag of socialism which, in various ways, had been part of its image history, with a red rose.[2] The new series

Fig. 1 An artificial fibre ICI tie promoting Fertilisers

of the company magazine now had a tagline 'for ICI people, worldwide', and claimed a circulation of 100,000.[3] In issue 5/1988 of *The Roundel*, a report noted that, as part of the corporate identity overhaul, the historic name Division was being dropped, in favour of simpler, more customer-friendly names. Plant Protection Division, for example, would be renamed ICI Agrochemicals.[4] In the same issue, there was a primer on 'Speciality Chemicals', the sector that ICI was developing. These were 'unique and different', with a 'customer service component'. ICI's portfolio included: 'biocides, polymer additives, surfactants, mining chemicals, polyols, silicones, Stahl's leather finishes, Converters' flexographic inks, Thoro System's cement and masonry sealers and coatings and Tribol's high-performance lubricants.'[5]

By 1989, two years after Harvey Jones's departure en route to becoming a TV celebrity businessman with the series *Troubleshooter*, ICI's profits were an amount in excess of £1.5 billion. A 1989 issue of *the Roundel*, reported on the company's fortunes.

> FINANCIAL RESULTS Profits before tax were up 12 per cent to £1,470m - which compares favourably with the large increase in profitability during '87 (29 per cent). Earnings per share went up from 113.6p to 129.7p, while the dividend increased from 41p to 50p - up 9p. The Chairman stressed two other points of strength: ICI's interest cover is now ten times, while the low level of the gearing ratio means that loans to capital employed stood at 26.8 per cent.
> MAJOR ACTIVITIES: In 1988, ICI made 30 acquisitions, costing £265m, and also 15 divestments, which realised £149m.
> The major acquisitions included: minority shareholding in C-I-L; 50 per cent holding in BAPCO in Canada; Williams Holdings, an Australian paints group; Thiele Engdahl, USA; and two Korean companies.
> Divestments included parts of Stauffer, Visqueen, silicones, 'Terram' geotextiles, and Coopers Animal Health. Capital expenditure was £811m for 1988, and included two projects in Japan (Hokota site and 'Melinex' plant), MMA capacity in Taiwan, a pharmaceutical plant at Severnside, and a fermentation plant at Billingham.
> Other significant developments included: the stock exchange listing in Tokyo, two new heart drugs, more MDI capacity in the USA, substantial new 'Melinex' capacity in Brazil, and a joint paints venture with Du Pont in Europe.[6]

98   E. LESLIE

Bulk chemicals still formed an important aspect of the business and were to be strengthened, the Chairman reported. Elsewhere in the same magazine were signs of what was to come. An article on 'Tigers' in the Asia–Pacific region, described new economic players, categorised as giants, tigers, rising stars and minnows. It reported on the thoughts of Robert Houston, former head of ICI Petrochemicals in Teesside and now chairman of the newly-formed ICI Asean Group, the acronym referring to the Association of Southeast Asian Nations, founded in Bangkok in 1967.

> Many, perhaps most, people believe that the Asia-Pacific region as a whole is already well on the way to becoming the world's new economic power-house for the twenty-first century,' he says, 'Between 1986 and 1987, GNP in the area grew by 6.3 per cent - over three times as fast as Western Europe and more than double the rate in the USA. This means that, by the year 2000, Asia's wealth should be nearing America's - and could have overtaken that of Europe.
>
> The mainspring of this is, of course, Japan, whose vast new wealth is transforming the whole region - which Japan regards as its home market. But, as the Yen goes on strengthening, the cost of manufacture in Japan itself goes up with it, so that their international business has to look further afield. This is most true of labour-intensive industries such as textiles and consumer electronics. South Korea and Taiwan, already well developed economically, are also rapidly following suit and the ripple effect continues to spread outwards. The pace of change is very fast, and several of the traditional agricultural economies in Asia are now poised on the threshold of industrialization.[7]

ICI's hope was as a deliverer of chemicals and intermediate materials to make the developing industries possible worldwide. In the company magazine, signs of a concern about environment and image in relation to environment, made themselves felt. Mike Flux, ICI Group environment adviser, wrote an opinion piece, 'Don't Expect Instant Fixes', in *The Roundel* 5/1988:

> The popular stereotype of the 'good guys' of the environmental movement ranged against the 'villains' of commerce and industry has always struck me as dangerously simplistic. For one thing, it downgrades the important contributions which industry makes to our everyday quality of life. Whether one thinks of our relative freedom from hunger, disease and drudgery, or of our ability to travel and broaden our education, I cannot

believe that these are aspects of modern life which most people would willingly surrender. Of course, none of this is to deny that our technology, and in particular the scale on which we now deploy that technology, has brought major environmental problems in its wake. But we should not develop massive guilt complexes about these problems unless we are failing to address them. Just as the need to improve nutrition, health and living conditions were a key part of the agenda for previous generations, so the need to develop new styles of industry and new lifestyles which are better fitted to life on a smallish planet represent the challenge both for us and for future generations.[8]

And the research was advancing at ICI into ecological modes. *The Roundel* 3/1990 announced trial testing at Billingham, of Biopol, a biodegradable plastic made from sugar fermentation.[9]

There was another future stuff being modelled and this was one in mould. A push for edible biomass stemmed from a panic about nutrition articulated in a paper from the United Nations in 1968. With echoes of William Crookes's warning in 1898, it argued there would be too many humans and not enough food by the 1980s and so artificial protein was needed.[10] Industry should cultivate the development of single-cell protein for direct consumption or animal feeding. As petroleum and chemical companies explored food from yeasts, bacteria and moulds, J. Arthur Rank, too, who had been in the cinema industry but turned to flour manufacturing, joined first with DuPont and, then with ICI, to find a way to make protein from his excess wheat. But, in 1967, so the story goes, he found a fungus that grew easily in fermenters and made a protein food. The Agricultural Division of ICI had developed the world's largest bioreactor—the 1.5 million litre Pruteen reactor, which was used for the cultivation of animal feed. Pruteen was the trade name given by ICI to the microbial protein produced by bacteria grown on methanol produced from methane or natural gas, as a replacement for all the soya-bean meal and fish-meal in the diet of sows. ICI scientists 'modified a bacterial metabolic pathway', by introducing a gene into the Pruteen molecule, which enabled it to fix ammonia efficiently.[11] High production costs and falling prices of competing products led to the abandonment of the animal feed project and the bioreactor was demolished in 1988. But adaptation of Pruteen as a human foodstuff, trademarked Quorn, was successful. The brand was approved for human consumption and launched in 1985 by Marlow Foods, as a joint venture between the bakery giant Rank Hovis McDougall and ICI. While mushrooms are the fleshy fungi that grow above the ground, beneath the earth, mycelia, root-like threads develop and these are mushed together to make the foodstuff. Quorn is a synthetic meat, ultra-processed, industrially-generated mud fungus, brought into being through uninterrupted aerobic cultivation of the filamentous fungi Fusarium venenatum under steady-state conditions. Massive fermenters jiggle the fungi—as they feed on a sugar solution made from wheat—in loops, centrifugally. The biomass is removed from the fermenter and injected with steam to kill off its cells, reducing the RNA, in order to make it acceptable for human consumption and not to risk the encouragement of gout and kidney stones. After four days, the substance is ready to be frozen, its long strands pressed tightly together, then the flavour is added and the meat analogies—mince, ham and chicken slices, sausages, meatballs, nuggets—derived. Quorn might be formed like chicken, bound with egg, flavoured, shaped. It was a minority taste and took until 1998 to make profits.

ICI began its fall into economic decline, as the effects of recomposition in the chemical arena worked through it. ICI made changes, reorganising its sectors, cutting back where it could. Tactics and strategies and organisational change caused profits to rise again, and shares climbed, though not to the peaks of a decade before. But the debts were hefty. The domestic industrial customer base had been rapidly retracting since the beginning of the 1980s. Competition was globalising. Burgeoning stricter environmental controls in certain regions of production dampened the ability to make profits and made closures or divestments a desirable strategy. A more selective focus was attempted under new leadership, restricting lines to seven sectors in the early 1990s: pharmaceuticals, agrochemicals and seeds, specialities, explosives, paints, materials and industrial chemicals. The company was seen as 'in trouble', and a corporate takeover company, Hanson, tried to move in on them, in a hostile takeover, buying, initially, 2.8%. Resistant to this takeover, ICI was pushed to more extreme measures to regain market share.

ICI contracted. It sold off parts of itself. It broke into bits. In 1988, the *Roundel* magazine devoted an article, titled 'Looking for a Parent', to what it saw as a well-managed divestment of the polythene film business Visqueen, in Stockton-on-Tees.[12] In 1991, agricultural and merchandising operations were sold off to Norsk Hydro. Its earliest business—soda ash production—was acquired by Brunner Mond, a descendant of the same firm who had in the very beginnings sold it to the company.

A demerger took place in 1993. *The Roundel* magazine, in 1992, devoted several pages to reporting on plans. A subheading, 'Splitting for Growth', made an analogy to strong plants in nature that split into two healthy plants capable of more growth.[13] The organisation was split in two. ICI held onto paints, materials, industrial chemicals, explosives and surfactants. The bioscience businesses, including pharmaceuticals, agrochemicals, specialities, seeds, speciality chemicals and biological products, and Marlow Foods, transferred to a new company called Zeneca. This later, in 1998, merged with Astra AB to form a company called AstraZeneca, a firm that became a household name after the global pandemic that began in 2020—it produced COVID-19 vaccines. The bioscience sector extended aspects of the company's traditional connections between divisions, given that areas of drugs and agroscience had over the years been cleaved together. Pharmaceuticals had been developed through a mingling of organic chemistry and biology—to cite two examples, one of the breast cancer treatments developed by ICI drew on agrochemical work on antifungal agents and, in a broader sense, the science of dyeing had been a stimulus for new drugs in the 1930s and 1940s. Zeneca was oriented towards high-costing bioscience products that fulfilled specific needs or produced particular effects. The chemical lines of the remaining part of ICI were more diverse in form—paints, explosives and many more plastics and other things—but all of it was to be made in large quantities and sold at cheap prices.

Still ICI kept on contracting. It sold off parts of itself. It broke into more bits. In July 1993 the nylon business was sold to DuPont. In March 1994, polypropylene operations were sold to BASF. In August 1994, its engineering services were sold to Redpath Engineering. In February 1995, Union Carbide bought the ethylene oxide and derivatives business. Some Unilever businesses and an American paint company were bought in the mid-1990s, as the shift away from commodity chemicals took hold—other types of speciality chemicals—higher growth, higher margin businesses—were acquired, leading to £4 billion of debt and the decision to dump more of the old divisions. In 1997, ICI Australia went. More and more sales went ahead. The fertilisers business was sold to Terra Industries. In 1998, polyester intermediates were sold to DuPont and Air Products bought methylamines and derivatives business. In 1998 Enron bought ICI Teesside Utilities and services business. In July 1999, Huntsman took on polyurethanes, titanium dioxide and some petrochemicals businesses. In October 1999, the acrylics business went to Ineos

Acrylics. In 2000, the methanol business was taken over by Methanex of Vancouver. In 2002, the Synetix catalyst business was sold.

Its fortunes were not saved. ICI withdrew increasingly from bulk chemicals and sold off more and more product lines. It also acquired products, as it shifted towards consumer-oriented products. Its paint line Dulux announced itself at the beginning as a fast drying, synthetic resin of unvarying high quality, elastic, never cracks or crazes.[14] It had remained a successful part of the company and so, attention at the company turned more to embellishing environments and bodies. ICI went into the manufacture of perfumes, cosmetics, artificial flavourings, acquiring, for example, Quest International from Unilever in 1997. Quest made flavours for snacks, chewing gum, dairy products. It made perfumes too. ICI could enhance the world through chemical consumption in more intimate environments. ICI acquired National Starch, which made food additives, such as those that keep corn flakes crispy, as well as adhesives and resins. From crunching up the landscape, swallowing the wind and the rain, to make explosives and fertiliser, ICI turned its chemical knowledge to another type of immediate environment, that of the self. The press-button age of the 1970s had shown the way: environments could be carried in a tin, or rather specific masking environments of air freshener could. Strawberry flavour could be imitated in three hundred ways. Something that claimed to be banana, but only ever tasted of artificial banana, could be synthesised. Perfumes and deodorants, hand creams and lip glosses could coat bodies and enhance, supplement or negate their natural smells and sheens. A synthetic world of imitation and deceit was at hand. It was necessary to work on the self-image too. The roundel of initials and waves was updated in 1998 by branding firm Corporate Graphics International. The various versions in circulation were unified in one single logo, a simplified version that retained the even, not choppy, sea lines.

To give some intellectual substance to their shift of direction towards the sensory, in 2002, ICI, under chief executive Brendan O'Neill, engaged an Oxford University academic, Charles Spence, to write a report on a new age, post-bulk chemical age of multi-perceptual settings. It was titled 'The ICI Report on the Secrets of the Senses: In association with Oxford University'. Brendan O'Neill provided the introduction, which laid out the transformation in ICI's sense of its mission:

It is by getting closer to the consumer that we are able to create imaginative solutions to the challenges of a modern lifestyle and to delight the senses. Over recent years ICI has changed, evolving from a more traditional chemicals business to a specialty business that stimulates the senses through the supply of ingredients for food, flavours, fragrances and personal care products, as well as decorative paints. ICI is increasingly being recognised as a vital ingredient in satisfying sensory needs.[15]

ICI-produced ingredients.[16] ICI was itself the ingredient, the spice of life, the icing on the cake of existence. The new approach of producing stimulating specialities was given the name 'Sensism'. Sensism was a 'science' that aimed to increase the capacity for pleasure, love and success. The brochure was full of highly colour photographs, of lovers kissing or caressing fingers on oily body parts, or girls and women smelling flowers and spinning tops a blur of red and rainbow, power-dressed women emotionally satisfied in their offices on telephones, blood red wine gushing into a glass. Sensory superpowers were the aim. The report concluded: 'Sensism is nothing short of breaking the code for the way we live'. The report diagnosed a lack, an existing sensory depletion, the result of an overemphasis on the visual: 'We live in a visually dominant society, which has marginalised the more emotional senses of touch and smell'. We long for touch, insisted the promulgators of Sensism. We needed more than the distanced senses of vision. We needed stimulation that entered the body or caressed its surfaces. We desired olfactory stimulation: molecules penetrating the nose. We needed soothing background music to shop to, soft tones seeping into our ears. We required motivating wall colours and coverings, which worked on our emotions in a subtle science of atmosphere. We needed to apply our scientific knowledge to our food's textures. Inside our mouth, its crunch and creaminess, its prickles and slurps would reanimate our senses. The report noted an emergent self-indulgence that was to be encouraged across society. This new, postmodern self-gratification, was one, the report stated, that did not take on the shape of 'hedonistic excess', but was rather 'an elemental and biological need, rooted deep within the body'.

Beyond the anhydrite locked in the ground and ready for making into commodity chemicals, ICI turned its attention to something intangible, a demand buried deep inside the human body. The elemental was in us, in our biology, not in the ground. It was in our senses, which longed for stimulants. It was there, but it needed to be supplemented, manipulated, chemically addressed. ICI would bring it up and out. Several times within its report, the phrase appeared: 'Sensism: a central biological necessity as crucial to our wellbeing as the air we breathe'. The air we breathed was not the whipping wind at the Teesside coast that was captured to make a chemical empire. It was an ideology that characterised experience in the postmodern epoch. Spence's research spoke of 'taking charge' of the senses, through ICI's new products. Food and drink featured prominently, for ICI's acquisition of Quest International was the context of the report. It described recent developments with the use of colour additives in relation to the experience of taste:

> Adding red to strawberry yoghurt, for example, will increase the perceived intensity of flavour even on the tongue of a trained taster. In turn, tasters can be fooled into thinking they are drinking red wine by the simple use of colouring in white wine.

> Colour can also create perceived sweetness, which is especially important for those on low sugar diets. Food scientists may soon be able to increase the pleasure we get from food and drink by carefully selecting the colour that best complements and enhances the taste and smell of particular products. Multisensory enhancement has a key role to play in the arenas of science, medicine and commerce. It will provide hope for those with diminished senses and an exciting way forward for the products of the future.

Supplementation could remediate lack. Even the experts could be tricked in this new synthetic science. Senses were elemental, but they could be deceived too, it seemed. A science of fakery and manipulation was in development. The report commented:

> Major scientific advances are being made in the quest to improve flavour through multisensory enhancement effects, as well as using visual dominance techniques to make foods taste better by looking better.
> Critically, the increased use of flavour-enhanced foods can lead to an increase in the palatability and intake of food, as well as improving appetite, immune status and other aspects of health and wellbeing.
> These endeavours are by no means only restricted to food. Making environments more stimulating for the elderly by using enhanced colour schemes, the increasing use of textiles, textured surfaces and other tactually

rich objects in the home can stimulate the sense of touch and make up for the effects of 'touch-hunger'.

There were other examples of how a focus on stimulation and staging within environments could affect outputs. Air freshener smells and lighting might be deployed to increase productivity. Perfumes could be used to increase our persuasiveness in crucial business discussions. Companies should turn attention to developing 'signature scents', in order to produce moods amenable to buying.

In these final years of ICI, the company seemed to draw closer to that which apparently represented its fantastical version, as conceived in its earliest days. Its proposals approximated the outputs of the factory as described by Aldous Huxley, after his time at ICI Billingham in 1931, shortly before publishing *Brave New World* in 1932. Huxley's vision of industrial organisation and production, was lorded over by a man named Mond, and everything revolved around consumerism, hedonism and moods enhanced by a drug called Soma. In one scene, the characters Savage and Lenina, who was prone to dress in 'bottle-green acetate cloth with green viscose fur at the cuffs and collar', and sat deeply sunk into their pneumatic stalls at the entertainment house.[17] The opening act is musically-based but nasally received:

> The scent organ was playing a delightfully refreshing Herbal Capriccio – rippling arpeggios of thyme and lavender, of rosemary, basil, myrtle, tarragon; a series of daring modulations through the spice keys into ambergris; and a slow return through sandalwood, camphor, cedar and newmown hay (with occasional subtle touches of discord – a whiff of kidney pudding, the faintest suspicion of pig's dung) back to the simple aromatics with which the piece began. The final blast of thyme died away; there was a round of applause; the lights went up. In the synthetic music machine the sound-track roll began to unwind. It was a trio for hyper-violin, super-cello and oboe-surrogate that now filled the air with its agreeable languor. Thirty or forty bars, and then, against this instrumental background, a much more than human voice began to warble; now throaty, now from the head, now hollow as a flute, now charged with yearning harmonics, it effortlessly passed from Gaspard's Forster's low record on the very frontiers of musical tone to a trilled bat-note high above the highest C to which (in 1770, at the Ducal opera of Parma, and to the astonishment of Mozart) Lucrezia Ajugari, alone of all the singers in history, once piercingly gave utterance.[18]

Nasal delight passed to aural wonder, all synthetic, all less than and more than human. The show took off, haptics, stereoscopy, the feelies and the scents continued to tickle all senses, to exceed what was commonly encountered—the highest note, a medley of delightful and disagreeable smells, mobilised symphonically.

Brave New World depicted what happened in this brave new world once the end of life of one individual arrived:

> It was a large room bright with sunshine and yellow paint, and containing twenty beds, all occupied. Linda was dying in company – in company and with all the modern conveniences. The air was continuously alive with gay synthetic melodies. At the foot of every bed, confronting its moribund occupant, was a television box. Television was left on, a running tap, from morning till night. Every quarter of an hour the prevailing perfume of the room was automatically changed. 'We try', explained the nurse, who had taken charge of the Savage at the door. 'We try to create a thoroughly pleasant atmosphere here – something between a first-class hotel and a feely-palace, if you take my meaning'.[19]

Synthetics allowed a passage into death, and the senses had to be stimulated until the end, so that some company could make profits until its consumers' dying breaths. Synthetics met the end of life and replaced life itself. Synthetics, electronic impulses, a manipulated atmosphere in a communal room, something like a cinema or a luxury rented room—an experience for everyone, issued by the factory from start to finish.

In this synthetic world, everything sapped the life and autonomy from individuals and from nature, in order to bestow on it a different nature, or a non-nature. Nature was remade from the nature that was there, in the rocks, in the air, in the sea. First nature becomes second nature. Nature has long been merged with second nature, which is a concept developed through the thought of Hegel, Marx and Lukács and indicated nature that had been augmented by historical and cultural intervention. Nature produced humans who in turn produced second nature. First nature, it might be argued, was a yearning, rather than a palpable thing. It was the idea of a nature that existed without human intervention or spoilage, a pristine nature, a wilderness. To see nature in order to call it 'first nature' was already to have humanised it within human designs. Second nature outspread. It encompassed all nature transformed by human activity—agricultural areas, urban constructions, national parks, river regeneration schemes, aspects of nature as tradable goods and resources hauled up from the earth and shaped industrially to make more things. With the successful dominion over the planet by busy capitalist resource extraction and the overspills of pollution as its result, it might be said that no place was unscathed by the effects of human activity: witness the melting poles or the plastic-filled oceans. In the ancient conception of nature of philosopher Marcus Tullius Cicero, first nature was wilderness, second nature was the sowing of corn, and both gave way to something he named third nature, which was aesthetic or strictly unnecessary from the point of view of survival. His example, the landscaping of gardens. Third nature was designed nature, and it proposed a magical or nightmarish world which came into being only as a result of infrastructural forms that were highly capitalised.

Highly designed nature is an art of synthetics. Highly designed nature, a development of threads through the synthetic and through the chemical, was what came into being out of the imagination of scientists, who might sometimes turn to science fiction to dream up futures. The developments forwarded by Sensism were already well in train at the time, utilising nanotechnology and other advanced forms to augment and enhance sensory perception in modes already envisaged in science fiction narratives. In 2002, an article titled 'Clothes that love you – textiles to touch the inner you' appeared in the journal *Pigment & Resin Technology*. It opened:

> Living in a world where vision is the dominant sense, we often neglect our other senses, like touch and smell. But, imagine a sense-enriched world where your sportswear cools you on the move. Your shirt shrugs off the smell of tobacco or repels mosquitoes. You sink into fields-of-flowers on your sofa, catch the smell of success in a boardroom or hit a 'cool' note in a chic hotel lobby.
>
> This can all become reality due to innovative new technology, which incorporates fragrance into textiles. Sensory Perception Technologies (SPT™) is the way to weave well-being back into our clothes and lives. SPT™ is the joint creation of leading sensory designers Quest International, a member of the ICI Group, and one of the world's most recognised textile brands, The Woolmark Company. This new technology weaves benefit-laden particles such as moisturiser, deodorant, fragrance, fresheners, repellents or even anti-tobacco agents into any fabric. It heralds an extraordinary new era of 'smart' clothes which soothe, stimulate, protect and cosset you! SPT™ even withstands washing machines, so its impact can be enjoyed wear after wear.[20]

Wellbeing could be woven into clothes. Particles could carry benefits. Clothes could be smart—smarter than they were, smarter than us. This third nature was better nature. This patented technology incorporated miniscule waterproof droplets into fabric, in 'microencapsulation', and these were activated by movement or touch. As we walked, we released something from our clothing.

Shibani Mohindra, the New Business Development Director at Quest International in 2003, explained the reasoning:

> Until now, textiles have only used the senses of vision and touch but a new way of looking at the impact of the senses on our lives – Sensism – will revolutionise the textile industry. This new philosophy is based on the idea that multi-sensory tools can empower us to become more productive, more successful and to enjoy relationships more. The first practical application is SPT™ which creates '3D fabrics'. The fabric, which will be the norm in the future, makes clothes that look good, feel good and do good.[21]

There was so much goodness to be attained in this utopia that was brought to us through the clothes that caressed our bodies. It extended into all manner of smartness around us, with air freshener and anti-bacterial substances emanating from carpets, as we walked, cosmetics exuding from cosmetotextiles, pesticides and flavours micro-sprayed or drugs delivered by nanoparticle transport systems. The world was to be monitored and augmented at the tiniest scales. We walked through our own little cloud environments, our own personal atmospheres tweaked and fine-tuned as we went. The best of all possible worlds was one for us alone. What of the bigger clouds—as ICI contracted, what happened to the clouds in Billingham? Did they contract too? Did the clouds disappear or did new clouds come, cloudy skies, no blue sky thinking, no blue horizons of the future?

The sell offs continued. In February 2006 the Uniqema business went to Croda International. The factories died slowly. Sensism did not save ICI. The financial press followed the moves. The *Financial Times* reported on 24 June 2007 about the scale of change.

> Some businesses have changed hands several times since divestment from ICI. Communities that once depended heavily on ICI have been through frequent and unsettling changes. For example, Teesside had ICI's biggest concentration of manufacturing assets, where 40 years ago the company employed 30,000 people and dominated the economic and social

landscape. Following a wave of divestments since the early 1990s that culminated in the sale last year of surfactants maker Uniqema to Croda, ICI's staff in the area now number barely 100. Nevertheless, there are still 13,000 people in Teesside who work for global petrochemical companies – as many as before ICI's big rundown began in the 1990s. Between them, those companies are considering £5bn of investment according to the North East Process Industries Cluster. Stan Higgins, chief executive of Nepic, says: 'This story grandad told of the dwindling chemicals industry is totally wrong ...this is a boom town.' In the late 1990s there was an anxious period as several owners of former ICI assets closed plants. The current mood is more positive – although competitive pressures are intense. Mr Higgins says there is a powerful argument that diversity, rather than unity, is strength. 'These businesses are generally more supported, more focused.'[22]

The market was free. The market was global. Capital took flight—it was free as a bird. It was global. It had a world stage. Other places were cheaper for manufacturers. Other places had features that facilitated the production that was demanded. Other states made offers too good to ignore. Teesside, in any case, moved from an economy centred on industrial production to one centred on services, leisure, retail and call centres. Between 1971 and 2008, in Teesside, 100,000 jobs in manufacturing disappeared, but 92,000 in the service economy appeared.[23] Everything was shifting. The story, the history, started to unravel. Everything changed, just as it always does, all the time. Or nothing changed, but it just went on rotting and inertia set in. But that too was a form of change. It was the change of decay—the counter-symbol to paints that never fade, dyes that never wash out, fabrics that never crease, polypropylene bowls that never crack or pall, ropes that never fray.

In time, the company was left largely with only Dulux paints, a rump of a firm which was taken over by AkzoNobel in 2008, its adhesives business transferred right away to Henkel. ICI faded away, like a stain that pales over time, or a dispersion, like clouds of chemicals wafting on the wind or exploded dust floating slowly away from itself.

The nuclear waste never came and the area lost its former uses, its purposes. It threatened to become more or less a wilderness, the name ascribed to it when Margaret Thatcher traversed it in 1987. The Prime Minister, dressed in a dark two-piece suit with matching handbag, picked her way through a wasteland that was Head Wrightson's engineering works in Thornaby, closed since 1984. She asked those who listened to her to imagine new futures that could take place here in what was to be dubbed 'The Venice of the North East', as the Teesside Development Corporation put it: those new futures were to be of business development parks, regenerated marinas, shopping centres and an expanded service sector. Might she have been able to dream as far ahead to freeports? She set the movement to the future in train and it came into being as her legacy. From her regime stemmed the commitment to enterprise zones, managed by unelected boards, existing outside local authority control, even if the local authority buys in and even if some members might, individually, be set to benefit. As an idea, it passed through Tony Blair and took on the name of regional development agencies, which proposed ambitious infrastructure projects, under the management of quangos such as ONE NorthEast (1999–2012), Tees Valley Regeneration (2002–2010) or Tees Valley Unlimited, set up in 2011.

The future to come of business development parks was one thread to be followed by various advocates, one lifeline out of all this decaying mess. Sensism was a phase passed through and abandoned. Charles Spence moved on to other commissions and new research. In various articles, including in the journal *Flavour*, in 2015, he directed his energies towards the higher-end molecular gastronomy fad, writing of 'Olfactory dining: designing for the dominant sense' and he highlighted the uses of atomisers, liquid nitrogen and dry ice—to make fragrant smoky mists and clouds—in cooking and cocktails and he considers the uses of nasal dilators as flavour enhancement devices and the technologies of 'inhalable food'. His thoughts turned futuristic:

> In the future, it would also seem likely that there may be a role for the hand-held technologies that one finds in most people's pockets in augmenting the dining experience too. Indeed, the last few years have seen something of an explosion of interest in digital plug-ins (e.g. the Scentee; https://scentee.com/). Certain of these can already release specific food aromas; just take the following: 'Available scents include rose, mint, curry, jasmine, cinnamon roll, lavender, apple, strawberry, ylang-ylang (a fragrant flower), coconut, and if you remember the fried corn soup fritters at KFC Japan from earlier this year, the corn soup scent should come as no surprise. There's also a limited-edition Korean BBQ collection with two meat scents and baked potato. Other scents are also in the works.' Currently the bubble-shaped Scentee retails at around $5 in The USA and can deliver around 100 bursts of a fragrance.[24]

If there was an impact on the senses, it came not from the doctrine of Sensism and its ambitious desires to manipulate at the molecular level the world of consumption. It came from a molecular transformation in the bodies of those who had been with ICI through a certain period of time, and some of whom suffered more than they had bargained for. The landscape of industry takes root inside the body. The labour environment had suffused some lungs with asbestos, the fibres that lodge in the lungs and steal the breath. A slow harm caused by asbestos mesothelioma, a type of cancer derived from the atmosphere, came into visibility. The legacies emerged only very gradually into the world: Asbestos fibres lodge inactive in lungs for up to half a century. These fibres travelled home from the factory. Here is one of many replicable stories:

> Joan Smith, 82, 'shook and wiped' the dust and fibres off her late husband Ronald's overalls three or four times a week for 30 years. She is now demanding of more than £200,000 in damages from his former employer after she contracted malignant mesothelioma.
>
> Mr Smith worked for ICI as an instrument artificer at the ICI Petrochemical Works in Billingham, where he was exposed to large amounts of asbestos through his work, his wife says. He died on August 28, 2000, aged 66.
>
> He often had to break off decayed asbestos with a hammer, and to work alongside laggers removing and replacing asbestos insulation, and his work clothes became coated with asbestos every day, according to a claim issued in London's High Court.
>
> Mrs Smith, of Rimswell Road, Stockton, washed and cleaned his clothes three or four times a week for 30 years, and shook and wiped the dust and fibres off the clothes before washing, she says.
>
> She put his clothes in a clothes basket before pulling them out to wash them, and these actions caused clouds of asbestos dust and fibres to be released into the air around her, which she inhaled.[25]

As the corporation disappeared, becoming absent, other presences, other remnants, other left behinds came more or less sluggishly into visibility. These joined the intangibles of memories and stories and past gestures made day after day in working lives and now no longer necessary. New imbroglios in legal and bureaucratic, media and medical institutions became apparent. Jim was just one who died a few months after feeling unwell, after decades away from the paints and specialty polymers and the asbestos dust. Many Teesside messenger boys, cycling through the asbestos-lagged large sheds to deliver their notes around the site, warmed up their daily lunch of tinned beans on steam pipes lagged with asbestos in the 1950s at the Billingham plant. Those fibres made their appearance palpable some forty years after. Others had laid out asbestos blankets to catch the sparks for welders and then, after use, would fold and take them into storage. The dust and fibres of asbestos had to be swept off the bare hands daily. In the Stockton plant, young men monitored the pressure and flow of ethylene gas in pipes lagged with crumbling asbestos. What legal claims against a company, now broken into parts, which had allowed this substance to interfere in lives? Others had worked in the 1960s and 1970s with benzene as well at the ICI North Tees Oil Refinery and succumbed, as a result of exposure, to myelodysplasia and mesothelioma. One man with asbestosis recalled crawling through ducts and around pipework so swathed in asbestos he was virtually eating it at times.[26]

The corporation had not only altered the landscape and made people. It was suffused through the landscape and people. Chemicals exist in combinations. To understand reactions means to realise what one thing does to another; how there is synthesis and change; how what exists is a result of what was once existing through time. ICI was a combination—a conglomerate, an expedient grouping to compete in a particular phase of internationally warring capitalism. The combination might be dismantled, but the effects of its presence, its long slow violence, are banished with more difficulty (see Fig. 2). The archaeological work of digging through layers, or uncovering, is costly, not always a priority. What happened in Teesside over decades was not unique. Around one of the IG Farben factories, in Wolfen, in the East of Germany, the chemical residues of colour production for Nazi films, processed by women, sometimes slave labour, in darkness, sunk deep into the ground. The factory was used by ORWO for film and photochemical production and fibres, during the years of the GDR, and today still, making niche black and white camera film. It sits atop a poison lake. When the city floods, it is said, chemical colours gush out and colour what they touch.[27] Industrial rainbows contaminate the land. What might yet come to light and might it harm us?

Fig. 2 The view from Wilton across the area

Grasping at Threads

What threads? What stories? What horror stories, or something that was not perhaps known, or not known by all, and which could not be known in its fullness until the event of its manifestation? What threads trail from past to present? In the preface to her book, *On Weaving*, from 1965, Anni Albers reflected on what it meant to weave:

> Though I am dealing in this book with long established facts and processes, still in exploring them, I feel on new ground. And just as it is possible to go from any place to any other, so also, starting from a defined and specialised field, can one arrive at a realization of ever-extending relationships.
>
> Thus tangential subjects come into view. The thoughts, however, can, I believe, be traced back to the event of a thread.[28]

The event of a thread. The threads included in Albers's life and imagination were those of a weaver who learned to develop her weaving art at the

Bauhaus in Germany, in the craft to which women tended to be restricted. She was an experimental weaver and she incorporated horsehair, cellophane and silver thread in her work. Part of her sensibility baulked at the move towards automatic weaving, for it limited her ability to react to a material. To weave is to produce an event. The threads of plastic materials in the factory, the threads of asbestos—these are far from the weaver's craft, but there is a thread from one to the other, a story of development, if not one of progress, always. Connections exist. Threads tangle with other threads. Threads allow for the telling of stories, for the placement of events next to events and on. All the mythical weavers brought a chain of events into being as they told their stories. Some chains are fatal chains.

What smells remained in Teesside? What sensory perceptions of the factories slipped into memories? Voices commenting online, in response to images in the Stockton picture archive, recall:

> In the 1950s when I was growing up in Billingham we often felt the blasting in the mine, crockery rattled and ornaments on the mantelpiece would shuffle about, does anybody else remember this?

Was this a collective experience above the ground, just as there was one below? Beneath the factory, men had stumbled, for eight hours a day, along narrow unlit tunnels, airless, uncomfortable, wrestling with machinery that, once it was expended, ended up in the Dead Man's area. And then came the periodic blasts rippling out from down there, waves expanding as circles across the environment, meddling with the arrangements in the kitchen and living rooms. The factory extends beyond its perimeter. The factory buries itself in us, in our senses, and also in our memory. What we hear and feel can be shared and confirmed or countered by others, especially now, as the logging and remembering of experience takes place through the various channels of the internet and does not go away again easily.

A thread on the Teesside Subreddit board about signs and signals from the factories around Middlesbrough speculated on the meaning of the sirens that blared across the cities. Reddit user Deanbbmv asked:

> What is that siren that goes off every now and then?

>> Usually it likes to go off as early as it can too, but occasionally it'll go off in the afternoon. Sometimes daily, sometimes it'll be weeks apart.

I've done some googling but all I can find is a non-existent post on the Gazette site. I assume it's from one of the chemical plants, which if so is it just a rather loud alarm for shift change, or is it a 'Warning people of Middlesbrough, this plant is soon about to explode. Think of the mutants for Futurama and you get the idea.'

ScubaCJ replied:

I'm in the sixth form at Macmillan Academy [Middlesbrough], and I hear it pretty much every tuesday morning at around 10am. I'm fairly sure it's one of the chemical plants testing the siren, but I could be wrong.
  One thing is for sure, hearing that siren on a foggy morning is scary as shit, it's like Silent Hill all over again.

Ashleypenny Parmo Eater contributed: 'haha I remember from when I used to live in Boro, it is something down the old ICI plants I think but I have no idea what it signifies!' Bethurz comments: 'I don't think anyone knows'. But Ashleypenny Parmo Eater retorts with: 'It's to signify that there's a fresh batch of Mutagen ready for any passing turtles'. The references to dystopian science fiction reinforced the historic links—Futurama, Silent Hill, experiments with DNA, are predictable in a place whose blazing chimneys piercing the night sky are said to have prompted the opening scene of Ridley Scott's *Blade Runner* (1982), and where Aldous Huxley's *Brave New World* (1932) was set and where, in the village of Carlton, Stockton-on-Tees in 1944–45, George Orwell's *Nineteen Eighty-Four* (1949) found its first form, as *The Last Man in Europe*.
  ColdFusion87 returned the conversation to something more factual:

It may be a daily test of the warning siren. I mean if there is a massive problem at one of the chemical plants they'd need to make sure the sirens are all fully functional!

Someone, now anonymous, offered: 'As far as I know it's something going off inside the Corus plant'. And the last word in the thread went to Panda_snizzle

I could even hear it in Stockton, but I still have no idea what it's for besides it's from the ICI/ Corus plants. When my bf (he's from Essex) first heard it he thought it was a WW3 siren... :P

On Casual/UK Reddit sub-board, a questioner from Teesside asked 'Does anyone else live in a place where WWII sirens go off every week?'

> It's attached to chemical/gas works so every Tuesday at 10am the sirens are tested as it will be used if there is a dangerous leak. It's only just occurred to me that hearing a WWII air raid siren weekly is probably not normal to 90% of people living in the UK

The responses tumbled in. Some were also from Teesside.

> Andythompson78: Fellow Teessider, when I hear the siren i assume its 10 O'clock on a Tuesday, if I then realised it ain't then panic sets in.

> bassviol: Norton School used to have three alarms. A fire one, an ICI chemical one and a third for escaped convicts from Holme house prison in Portrack 😁

Another thread on a paranormal subreddit board wove together local knowledge, myth, rumour, conspiracy and a downbeat humour about the past. Foggypatroller opened with an anecdote about parking up by an old church in Aisleby in Teesside one day and seeing a gravestone lit up by a 'lime green coloured mist'. His dad, too, in the 1950s, saw a spaceship the size of the town, but close to the stars, with ladders or rails and watery green lights that flashed off once the rails were full with round things. A reply came in.

> ledgerdemaine: 'That Green mist was probably ICI fallout. And aliens could have been from Middlesboro as well. Two heads I heard.'
> foggypatroller: omg how did you know about ici where abouts are you from „im a smoggy , thats what were called we glow up in the dark like readybrek kids ,fkn hilarious what you said „,sometimes i type on here and no one understands me as im also sometimes called a geordie even tho im in middlesbrough well stockton on tees but ppl know boro more„thanks friend xxx
> ledgerdemaine: Hey did you know Ridley Scott filmed the movie Aliens on beaches near there to look like an alien world. I live in Oz but am from Sunderland. My Dad lived in Stockton til he died. I know what you mean about being called Geordie, really annoys makems too lol. Used to work in markets at Darlo, Stockton and Middlesbro way back in 60s so remember ICI

> foggypatroller: he went to richard hind school with my dad,,on bowie lane,,,also him out the harry potter movie ,harry potters step dad,,,he only ever spoke in drama class,,,i cant believe you lived local....ici is shut now,,i did the darlo markets n stockton n boro ones but in the 90s,,,how random is that ,,good to hear from you

On a generic AskReddit board, a user introduced herself with the local nickname, Smoggy, suggested that the heavy air of smoke and chemicals had not only passed into the inhabitants' lungs, but had also overcome them fully. Smoggy referenced the idea that the region made its inhabitants radioactive and turned them into aliens in their own home.

> Accomplished_road484: Stockton on tees I'm a smoggy lol.we glow up in the dark from ici chemical place. Middlesbrough's football team is quite ok. They're a couple of mile up

The web, Facebook local history group pages, Discord servers, Instagram and Twitter all produce a shower of memories and queries about what life is, was and might be amongst the chemical industry and its remains and ruins. There is thread after thread. As many threads as were made from plastic pellets in Teesside? Nekogyph's Flickr feed displayed a photograph from the Dorman Museum, Middlesbrough, titled 'ICI Threads'. It showed four spools of synthetic threads, two red, one black and one blue.[29] They were to the side of a case that also contained a spoon, a plastic strip and two leaflets. One was a brochure for a Style 44 Winding Machine, which coiled the thread onto bobbins at high speed. The other was an explanation through diagrams of 'The Manufacture of Synthetic Filament Yarns and Stable Fibre', with its specification of the processes of melt spinning and drawing and processing. Behind it all was placed a slightly creased and neatly folded white overall with the ICI logo in black on the left side, just above the heart. In the comments on the photograph, Bolckow noted: 'My dad's overall:)'. This is a company overall and so it is generic and not unique—as is the essence of the uniform. But this overall too is a specific one, worn by a man, who had a child and who passed through a particular place and time and made that history and shaped that world.

Threads are made, and more threads come, but these now are the threads on the web. Threads about threads. Yarns about yarns. Just as it is possible to find—and laugh about in comments that deprecate ourselves and our ancestors—thousands of advertisements held in various online repositories that preserve evidence of past dreams and hopes and promises peddled by ICI: 'The more Bri-Nylon socks he has, the brighter he will be!', 'Fashion's more fun in "Crimplene"', 'Get in the picture with ICI Fibres', 'Tough. Elegant. "Terylene"!', 'Keep your child safer-from-fire in Bri Nylon', 'Cosak: Spells action', 'Only "Crimplene" gives twosomes such versatility', 'Burton forecast a blazing summer in "Crimplene"'. These synthetic yarns are staged in boldly coloured advertisements that found a home in publications such as *The Sunday Times Colour Section*, which was first published in February 1962. Glossy magazine colour and the use of photography accorded with the bold colours of the new synthetic fibres. The advertisements' rhetoric of sunny days to come and passionate lives lived in polyester fashions—in baby doll nighties and drip-dry suits—persuaded those who had the money and energy that good things could come through 'the white heat of technology', as Prime Minister Harold Wilson's speech, 'Labour's Plan for Science', given at the Labour Party Conference, in Scarborough, on 1 October 1963, phrased it.

We can pick up the threads of what was, find many threads that lead to stories and questions and answers and speculations about what happened in the chemical factories and around them. History frays into many threads. Multiple tellers become like the mythical weavers Arachne, Athena, Ananke, Philomela, Penelope, Isis, Neith, Tayet, The Norns, Frigg, Saulė, Holda, Spider Woman, Amaterasu, Mokosh, Habetrot. The filaments that they spin together and tangle up make worlds, fasten fates, tell stories, make histories. The world is made in stories. History is woven in stories.

News About Clouds

What clouds might yet be made? How might the senses still be stimulated by the remnants and transformations of ICI's factories? In 2013, orange clouds appeared above homes, a bright smoke spreading from Billingham Manufacturing Plant, result of a pressure release of Nitrogen Dioxide from the fertiliser plant GrowHow. Strange plumes were signs of a presence still in the area of things made, of nature in the process of being remade and reworked.

When the cooling towers of the ICI Lackenby Nylon plant at Wilton were demolished on 22 April 2012 at 11:30 in the morning, they uttered from themselves their final dense white clouds, spreading first from the base and, then, forced up through the crumbling tower itself. These drifted together into one great exiting cloud.

Industry still exists in the old factories or on the old sites of ICI in Teesside. Industry has not gone away, but its forms change and it employs fewer people and so less is known about what it does and less can be understood about how it might be shaping lives in its environs. Clouds still pump from chimneys. Photographic agency Alamy Stock has many high-resolution photographs of ICI Teesside. The irony is that the high resolution shows largely a detailed rendition of blur. Where is the focus when there is nothing to focus on?

New industries such as hydrogen production were proposed for the former sites. While ICI devolved down to AstraZeneca, other firms moved into the area to deliver biotechnologies and biologics. Fujifilm Diosynth Biotechnologies in mid-2021 expanded its production in Billingham of the Novavax coronavirus vaccine. But the billowing clouds, the fish smells, the pumping plumes of industry smog seemed like intangible phantoms from another age. The wilderness became a nature reserve, in part. Clouds of the past came and scurried away, compromised by the factory-smoke clouds, the pools of steam and the smog that hung heavy on the air. When these were fewer—through lack of industrial activity—it was not as if the usual old clouds came back. Something else scuttled in, some synthesis of cloud as a nature-cloud in historical guise.

Local newspapers, in November 2015, reported on the appearance of thousands of starlings flocking to make an immense dark cloud over the wildlife park at Saltholme between Hartlepool and Billingham. The murmuration gathered and swooped, ringed by cooling towers and chimneys, power stations, oil refineries, incinerator plants and dockyards, where toxic—asbestos-filled—shipwrecks were broken up, and the towers of Quorn production loomed. The birds arrived in small groups, gradually amassing, and slowly the clusters built up to a swirling, pulsing whole that shifted and coiled fluidly. It was a cloud of birds twirling in unison, before dropping into the reed beds, as darkness fell. The site where the birds gathered was a salt marsh that was once extracted industrially to produce caustic soda, and so inaugurated the chemical industry in the area. It came to be one of ICI's sites. The area became contaminated over the years—clay capped the poisons, keeping them buried in the ground. The water, the salt and the wilderness here make sense of the place—for both industrial excavators who will build an industry to change the world and the birds who wish to roost here for the winter. The water, the salt, the wind, the air, the swirling birds, the old nature making lives amidst the new nature, the old industry and the new: all this, as a visitor comment on the RSPB Saltholme website put it, was 'truly is a feast for the eyes and the senses'.[30]

Turn the focus to the skies above the old industrial areas. Repeatedly the local paper, the *Chronicle* and other regional newspapers reported on strange cloud formations, seen to accompany the increasing number of odd weather events. These clouds appeared for a few moments and scurried on—but left pixel traces on smartphones and on the web. In January 2017, the *Chronicle* asked readers to send in photographs of strange cloud formations. Laura from Gateshead saw Donald Trump in the clouds. Other faces hazed into sight. The British Isles appeared above the actual one. And there were countless striated skies and jagged wispy clouds and thick foamy clouds and clouds like footprints. In February 2017, a photographer caught repeated pink horseshoes in the sky. In 2020, a Marton woman saw strange blanket-like low clouds and a green object—not visible to the naked eye—in the sky.

In April 2022, under the headline 'Did you see it? "Crazy" cloud formation spotted in Teesside sky by family', *Teesside Live* reported on a 'strange cloud' above New Marske. It was tube-like and had something of the look of a tornado swirl. What impending doom did the cloud signal? The cloud came as the family was watching the film *Ghostbusters*. The weirdness of the cloud compounded the terror already being experienced by the eight-year-old child.[31] Perhaps, speculated the mother, such 'weird sky markings may link to planes flying overhead'. She 'scoured' the internet and a possible name for the phenomenon was identified: a funnel cloud. These come from cumulonimbus clouds—thunder clouds—thick with rain or hail. They were spinning fingers of cloud, or dangling ropes of cloud, on their way to make contact with the ground. A tornado on land or a waterspout on the sea would result and raging winds would sweep across the area. But it stayed there, in 2022, more a portent of troubling things that might yet be to come.

Were the clouds changing or were we just more alert to them? A new cloud was named in 2015 by the official body of cloud identifiers and labellers, the World Meteorological Organisation, and it was added to the International Cloud Atlas: Asperitas.[32] This cloud formation looks like undulating ocean waves projected into the sky. Following the tradition of cloud nomenclature, its name is from the Latin 'aspero', to make rough or uneven. Ernst Bloch wrote, in 1959, of a sea in the sky, as projected by the daydreaming mind or the fairy-tale gaze of a child. A cloud in a blue sky becomes an 'island in the sea of heaven or a ship, and the blue skies on which it sails resemble the ocean'.[33] The airy blue sky swaps out elements to become imagination's watery sea. Other daydreaming minds see mountains in the sky—air becomes rock:

> the distant mountains first appear, a towering and wonderful foreign land above our heads. Children believe that white-capped clouds are ice mountains as though Switzerland were up in the sky. There are castles there, too, taller than they are on earth, of ample height.[34]

Air, water, earth, rock: each transforming into each other. The sky acted as a mirror, reflecting for the benefit of fantasy earth's inversion. If down below on the ground there was the world of body and action and suffering, up there above in the clouds, for Bloch, was the world of mind, thought, imagination and other histories, including better ones. Above was a faraway realm and it tempted an earthbound gazer with the exhortation to escape this one. The sea in the sky was not limited by earthly coasts, in the fantastical view. There was a gigantic water mass up there. But here, around the new clouds, there is a troubled sea, not a blue one. The ocean is choppy, swirling and fatalistic.

Another newly named cloud, Fluctus, appeared. It was described as curls or breaking waves on the top surface of the cloud. Another one was named the Cavum, and it described a circular hole of sky peering through a thin layer of supercooled water droplet cloud. This spread out gradually, over time, dissolving the cloud around it, making the sky gape, like something pitted or torn, with little wisps of cirrus or virga, drooping from its edges like fraying materials or disintegrating substance. Aeroplanes can make these clouds and they may drag them into lines across the sky.

New clouds resulting from human activity crowd the cloud atlas. Homogenitus is the appellation and it covers aircraft condensation trails or contrails, as well as clouds that come from industrial processes, such as rising thermals above power station cooling towers. And should the contrails persist for more than ten minutes they are given the genus Cirrus and the full name becomes Cirrus homogenitus. And so nature becomes humanised. These contrails that hang around over longer periods may be blown by strong upper winds across the sky, such that the human clouds become indistinguishable from the ones that formed before aeroplanes were in the world and so they need a different name, should one still know from whence they came: perhaps Cirrus floccus homomutatus or Cirrus fibratus homomutatus.

Birds roost under these skies, under new and ancient clouds (see Fig. 3). The clouds replace the clouds pumped from the chimneys and the cooling towers into these skies, these watery and wind-filled skies. Middlesbrough was called an 'infant Hercules'. Half-man, half-god, he suffered in his mythic life, willingly, to show off—and to test—his extraordinary strength and resilience. His legendary nature meant that a constellation was named in his honour. Hercules was one of forty-eight constellations listed by Ptolemy, an astronomer from the second century. Hercules, in the third quadrant of the northern hemisphere and visible at latitudes between +90° and −50°, is described, in a student's guide from 1891, as organised around a fine double star that is orange and emerald-green and its light variable: 'It may be just detected with the naked eye on a clear night'.[35] It is one of nearly a thousand star patterns observed today. Its stars are arranged in a diamond shape, and at each angled corner an articulated leg branches off. It could appear like a four-legged spider, or, at an imaginative stretch, as the Kneeling Giant, which it is also called. But it seems too much like a model of chemical bonds, as they appear in school chemistry textbooks. In that guise, Hercules speaks to all the chemical bonding and rearranging that took place on the Earth below it.

Nothing is fixed but the stars. But this industrial episode, this passage through the time of ICI and everything it brought into being and its aftermath, might tell us otherwise. Even the stars are not fixed. What we can see in them will change as we change. And whether or not we can see them will change as we change. In 2019, newspapers reported on a starry constellation newly visible above Middlesbrough after industry of a certain type had declined and the pollution and smog shifted away. This constellation of stars, of course, comes to be named not mythically, as is familiar from the ancient ones and might fit a city that was once called Hercules, but, instead, much more popularly, after the local high-calorie signature dish, Parmo and Chips.[36]

Owen Hatherley visited ICI Wilton for his study of the state of things in the UK, titled *A New Kind of Bleak*, from 2012. Though hit badly by recession, part of the site was still operating, as was obvious to his senses— 'the smell makes it clear that somewhere petroleum is being transformed here into something nature wouldn't want us to do, and that at least is

Fig. 3 Gulls on the beach at Marske by the Sea in February 2022

mildly comforting, especially when fully familiar, as I am, with the safety procedures'.[37] The building was modernist, with red brick wings and green plantings and a Bauhaus-style road, hauled up on pilotis, running beneath the building 'put there to create a feeling of motorcity modernity'. But it was dying its slow death. Or at least being forced, apparently, to grow up. He reflected on how the prime minister of the day, David Cameron, talked

> about 'weaning' this place from the teats of the state. Either way, the people who live here are treated as children. What is especially noticeable in Teesside, though, is that this 'public sector' has spent much of the last two decades trying to prop up, resuscitate, or bring into being a moribund or dead 'private sector' - regeneration companies and the sell-off of public assets to prompt property development, a new University to stimulate the 'knowledge economy', the building of art galleries to attract 'creative capital' and of 'shopping malls to inculcate consumerism. The public/private divide never looked so false as it does here, where the 'public sector' has long worked doggedly for the private, thus far without obvious reward.[38]

The idiom—the one spoken in the press and on the ground, the one that threads its way into lives, beyond the endless talk of employment opportunities and restructurings, reskilling and inward investment—is crowd-pleasing. As above, so below: Parmo and Chips. Humour is necessary in these super-serious days. Everything might still have been to play for. Everything was still being made or unmade. Some effects last beyond a lifetime. Some have disappeared already. The ground was contaminated. What was ICI? What will it have been? What is to be made of the remainders? What senses did it cultivate for those who lived amongst it and who live within its traces? What senses yet to harness? What sense can be made of it all? What lessons were learnt? What patterns were established and will repeat?

Here we arrive, close to now, when the future advertises itself as smart, smarter than we need to be, and the cloud as digital. A new age insists on clean technologies, on skies not earths, on renewables and sustainability not resource extraction, on locality not worlds away—that is the rhetoric. The new industry of the region focuses on biomass, biofuel, bioethanol and low-carbon energy derived from waste plants. Here the waste of the world will yield more power from itself. Here the atmosphere will be harnessed again, but this time for renewable energy assets, in particular offshore wind. The highest average wind speeds in Europe, battering the land from the northern North Sea, will be captured in the wind farms springing up around the port infrastructure (see Fig. 4).

Toxic Ground

On 22 June 2022, in Redcar, the coke ovens of the steel factory, which had been closed since 2015, were caught, from multiple angles, by drone footage, commissioned by demolition company Thompsons of Prudhoe, as they disappeared in what the local press called a 'mushroom cloud of dust'. A rip of sound and all at once, a grey cloud emerged at the base of several structures, rapidly swelled, as everything was dragged down, as if into the earth, as if a hole were opening up on the surface of the earth. The several clouds at the base of each building—a junction house, conveyors, including two 85-metre chimneys that had puffed some much out of themselves—became one cloud. The local newspaper reported on it, noting how, in a moment, a set of installations 'crumple like a piece of paper': 'They were once a daily sight for thousands of Teesside workers but within seconds, they were gone—obscured by a thick curtain of grey dirt.'[39] What had persisted, since 1979, chambers where coal was heated for 17 hours to make coke to feed the blast furnace, what had tapped into the long time of coal formation, the promise of jobs for life and day-after-day consistency, vanished in the blink of an eye. In the video footage, the dust gathered and greyed out the visual field of the screen. What was visible, tangible, was now opaque, over. The air was thick and the air quality monitors spiked, fizzing with the invisibly small, most toxic particulates. A furry beige cloud undulated over the vast site, pushed slowly as a mass in one direction. It crawled along the ground. It did not take off into the sky. It was here on earth.

And what should cover a now flat and scorched earth? New developments that promise, yet again, employment, opportunities, profits and a future on sale, under the auspices of this year's mode of modernisation—clean energy, industries that mop up after themselves. For all its promises, the older waves left so many poisons behind, sunk into ground and sea bed—and they left memories, collective or otherwise.

On 4 May 2017, the first Mayor of the Tees Valley was elected. Ben Houchen, a Conservative, was elevated into power by a small percentage of the eligible electorate, namely 21.31%, comprising the amount of those who cared to vote in order to make a point about how abandoned they felt amidst the wrecks of industry, work, nature and infrastructure. He was

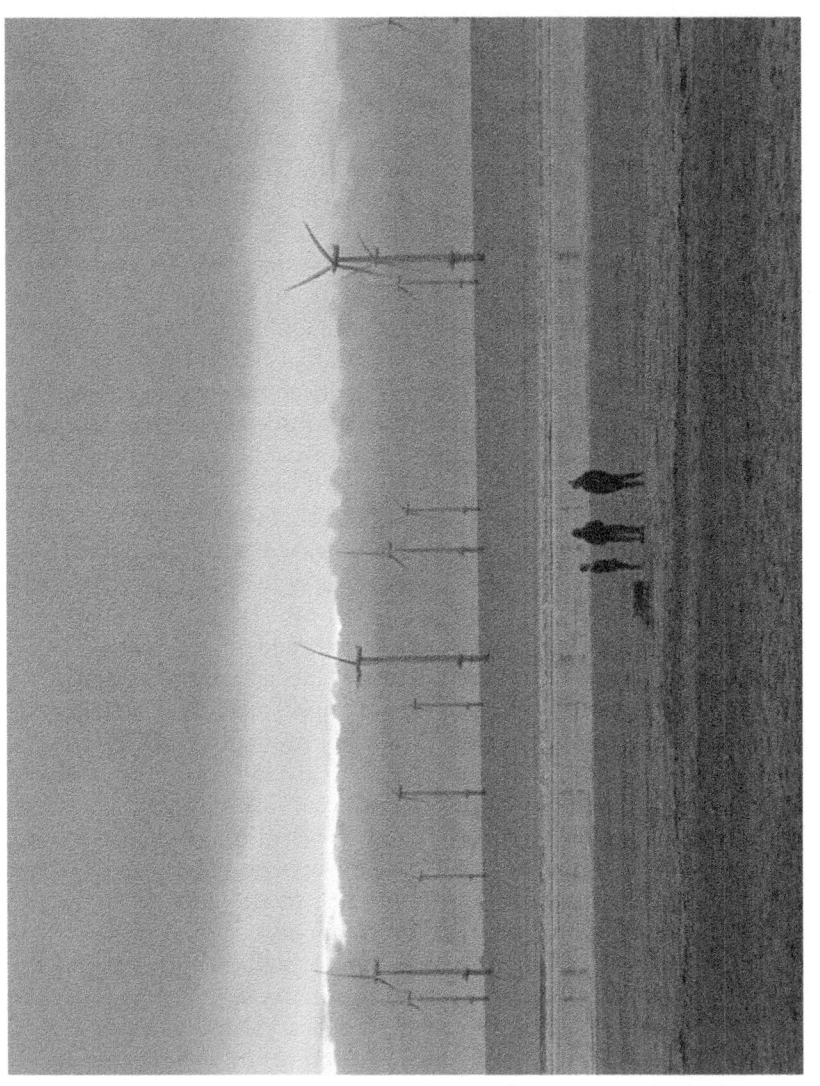

Fig. 4 Wind farm off the coast of Redcar in January 2020

reaffirmed in the role on 6 May 2021, with a larger mandate: 121,964 votes from a turnout of 33.95%.[40] All this undermined the Labour Party, which had for many years, since the Second World War, been a dominant force in the region, claiming to be the representative of working people, of trade unionists. Other political players had come in, anti-immigrant and anti-establishment, the establishment being those in London, who ruled, middle classes whose allegiances were seen to be elsewhere, far away from here, Political stances revolved around the relationship to Europe and its customs union. One area of Teesside—Brambles and Thorntree—is said to have collected the proportionally highest vote to leave the European Union in Britain when the referendum was held in June 2016.[41] They imagined in doing this they, as the often-repeated rhetoric put it, would take back control, get their lives back, get their money back, get their country back and get the chance to govern themselves. Get back this toxic land that is deposited with the rotten offshoots of their once proud labours: iron and steel works, coking works, railways, tar macadam, slag works, brine works, cement manufacture, anhydrite mining, landfill and chemical works and all of its infrastructure of buildings, plants, production facilities, pavements, services and waste storage and transfer areas.

They got a Masterplan for a new Teesside. The old factories would become the ground of a new economic region. The whole region was to come alive to the sounds of successful industry, emanating in and around the Teesworks and the Freeport, an expanded site of the old Teesport, covering 4500 acres and into which is folded Teesside International Airport, the Port of Hartlepool and the Wilton International industrial site. Developing the freeport was part of a flagship policy of the Conservative government.[42] Imminent arrivals on this land were the new industries inhabiting the Teesworks installations. Teesworks announced itself, on websites that touted for investment, as 'the UK's largest and most connected industrial zone'. It is a redevelopment of Europe's largest brownfield site. Security fences and guard dogs patrol the sensitive spots. These are internal borders around special economic zones. In development here in Teesside are compounds where tax and customs regulations are lifted, and business is incentivised with various exemptions, while planning permissions are made laxer. This is the future. Or this is a version of the future that may come into being. Many forces are gathered to make it real.

The Teesworks Freeport zone began as an entity held in public hands as much as private, but over time transferred largely into private ownership, once it had seemingly absorbed tens of millions of pounds of public money. What happened, what *is* happening, is murky, as murky as the mud-filled water dredged up along the river and in the sea, in the course of preparing the new ground (see Fig. 5).[43]

The old sites for the housing of the brave new world of industry are toxic. There are hazards for the aquatic environment and fiery risks from fire water, full of pollutants melted from the buildings whose flames they quenched. Over the years stuff dispersed into sewers, freshwater, estuarine waters, coastal waters, land or groundwater. To decontaminate these sites, to cleanse this dirty water, was not considered financially viable and did not feature in the master plan.

> The remediation strategy will be based on a do minimum/do necessary approach, to an end-user specification, and be one where there is flexibility in the redevelopment strategy to arrange end user site allocations to minimise conflict with localised, more heavily contaminated areas wherever possible. The remediation strategy is, therefore, not to create a blank canvass for development, where any future development scenario is permissible, but to take a balanced approach to remediation, applying innovative

Fig. 5 Sea off Redcar, February 2022

techniques and solutions to mitigate cost, optimise development configuration and, ultimately, realise a higher level of project viability. Importantly, the strategy will be geared to achieving earliest possible response times on the release of land in line with firm developer interest, so that revenue can be realised as soon as practically possible, working to an overall, co-ordinated vision.[44]

Instead of making clean again, a strategy developed around breaking chains of contact between harmful substances and receptors, the people, the water, the animals, the environment. Can each be sealed from each other? Can something be capped and so lie inert, under the landscaping and buildings of what comes next, awaiting only the catastrophic, unforeseen conditions of its release?

Here was a realm where anything could happen if capital commanded it, as had been, and still was, the case at the Wilton site. An Instrument of Consent, dating from 1946 and granted to ICI, meant that the sites could be used for anything without planning permission. And the political structures welcomed all with open arms, and the tangle of red tape of Brussels was snipped through, loosening up what can be done, what needed to be protected and what could be imagined.

The rapidly growing electric vehicle battery market is one area slated for development, an attractive draw for direct foreign investment, as more and more Lithium Hydroxide and Carbonate are required year on year and gigafactories spring up across Europe. Processing plants for the necessary lithium are proposed at Wilton International Plug & Play Chemicals Park. The largest lithium mines in Australia are to be linked to Teesside. It comes in from far away. Around here a lot that could be taken has been taken. There is still matter left. What is extracted now includes crushed rock and sand and gravel, with a little brine extraction at Seal Sands and some clay at Cowper Bewley. Underneath Wilton, there is brine, but the caverns beneath are used now for storage of gas. There is natural gas at Kirkleatham from a Permian limestone reservoir. What remains of the anhydride beneath Billingham could yet be mined. All this is on offer to those with the tools and the will and the money. But waste persists, and it becomes a viable industry. A massive urban quarry, a giant wash plant appears, sponsored by Tees Valley Authority Combined development grants, as the region heads for a Net Zero carbon future, to turn waste into products—pipe beds, bricks from aggregate diverted from landfill—for the building trade. Elsewhere in the site an aviation 'green fuel' plant is proposed to come into being, deploying a waste-to-liquid process to convert household and commercial waste into aviation fuel and Naphtha. And underpinning all this is the development of carbon capture utilisation and storage technology. Techniques are found to capture carbon dioxide from activities such as power generation or industrial manufacture using biomass or fossil fuels that emit it at high levels. Snatched from the processes, fixed from the air, it is compressed and transported through pipelines, on ships, rail or trucks to be used again or stored underground for the rest of time or until it bursts out again. In Teesside, the carbon dioxide is to be stored in the North Sea. Here in the briny swirl, off shore from the East Coast Cluster of Teesside and Humber, is a facility 'ready to remove almost 50% of the UK's total industrial cluster CO_2 emissions'.[45] The carbon dioxide is to be held in a saline aquifer called the Endurance Reservoir, ninety miles from the coast and just less than a mile below the bed of the North Sea. The sites are planned to expand until a billion tonnes is stored there. Is the carbon gone or does it just wait for another time? Is this a way of extending the time of the fossil fuel era, a seeming big win for those who promise to restore the earth, but, in actuality, nothing more than puffs of hot air suffused with greenwash ideology and fuelled by grants and the hope of profits, and accepted because of the promise of jobs to come?

And sometimes those jobs, those new industries, hurt. On 18 November 2021, the local press reported on a fire at a biomass factory in Redcar, whose smoke stank. It burned for days: 'People said the stink from the smoke was "all they can taste and smell" and it was "making them feel sick".'[46]

Another Conservative Party buzzword policy was 'Levelling Up', a policy designed to 'end the geographical inequality which is such a striking feature of the UK', by 'unleashing the power of the private sector to unlock jobs and opportunity for all'.[47] On 4 October 2021, a minister, Michael Gove, spoke at the Conservative Party conference, on Conservative Party policy:

> Levelling up means four things. We want to strengthen local leadership to drive REAL change. We will raise living standards, especially where they are lower. We will improve public services, especially where they are weaker. And we will give people the resources necessary to enhance the PRIDE they feel in the place they live.

And if you want to see all four in action and see levelling-up in reality, come to Tees-side. The Conservative mayor exemplifies great local leadership. He is bringing tens of thousands of new, high-paying jobs to the area to give many more a better life. He is changing the face of transport and further education to ensure services work better for all and especially those who've been overlooked in the past. And he's changing the face of Darlington, Stockton, Middlesbrough, Redcar and Hartlepool with new businesses and brighter high streets giving those towns their pride back. Tees-side neglected for decades under Labour, its proud heart nearly broken, now revived, regenerated, renewed by the CONSERVATIVE LEADERSHIP of the amazing BEN HOUCHEN! That is levelling up in ACTION. That is our party's mission for the WHOLE country.

As Michael Gove delivered his speech, in October 2021, crabs and lobsters, crustaceans that feed from the sea bed, washed up in large numbers onto the beaches along the North East Coast. Tossed onto beaches, they staggered and twitched, blowing bubbles from their gills, and, after a while, they collapsed in heaps, paralysed, before slowly life left them. Shellfish pickers arrived and, it is rumoured, the polluted crabs found their way into the finest restaurants. The heaped masses of lifeless crustaceans appeared like a negative counterpoint to gathered birds swarming around industrial pipes. The deaths continued in the following months, and into the following Spring. There were arguments as to the cause. Was it from pyridine, a chemical toxic to crustaceans and with a long history of release into the waters around the factories, from coking for steel, for example? A steel factory had been exploded just prior to a mass die-off, its cloud of toxins uncontained, floating freely, just at the same time as the Tees estuary was being dredged for ten days, all day and all night, with 148,000 tonnes of sediment removed and dumped at doses. Pyridine had long built up in the area, for example, discharged from an ICI plant as a by-product from the thermal cracking of Naphtha. Had it been dragged up from its rest in the river and sea bed by preparations for the regeneration of the area? Or was the cause of death a naturally occurring algae bloom?[48] No clear factor was found or stated as cause. Some novel pathogen was fingered, but it remained unnamed. The experts, one from a large Port Group, Peel Port, stated to the government that it could pass no decisive verdict, but the dredging was an unlikely cause. The case was closed by the government.

The assessment concluded, after considering several issues (including the lack of mortality in any other species) that releases of chemicals from dredging, an algal bloom or contamination from the chemical pyridine, as suggested by a study at Newcastle University, were not likely to be the cause of the deaths.[49]

No more investigations needed to be undertaken.

> The Environment Food and Rural Affairs Committee wrote to the Secretary of State Thérèse Coffey MP on 31 January 2023 asking her to refer the matter to the Centre for Environment, Fisheries and Aquaculture Science (Cefas) for further study. The Secretary of State did not agree, highlighting no long-term impact on crab and lobster landings in the area, and that:
> 'Given the extent of the analytical work already undertaken, and further advice, I have decided that it is highly unlikely that we will find the cause and so no further analysis will be undertaken by the government.'[50]

The dredging continued. In September 2022, it was necessary to dredge, in order to construct a heavy lift quay for the Freeport. The first dredging is called capital dredging, picking up sediments, debris and whatever else is found on the sea bed. Vast quantities of material were excavated and disposed of, the company said, on land. The second phase was to drop dredged materials at sea. Who could speak out against this? To speak is to be cast as outside rationality by those in power. In an interview for the Today programme on BBC Radio 4, the Tees Valley Mayor mocked those who called for the local authority to test for chemicals:

> It's not just pyridine, they think it's Agent Orange apparently from secret factories in the Second World War. We've also been told that it's Russian submarines trying to cause problems for the UK government. I'm sure you're not suggesting, and they are suggesting, that we do testing for these types of completely conspiratorial ideas, because if you do that we'll never get this development underway and finished and that's equally as damaging to the local people who in our local area want jobs and they want money in their pocket to look after themselves and their families.[51]

Perhaps something toxic welled up as a by-product from the past. Will it ever be known? Perhaps the toxicity resulted from chemicals made during the period of intense production in the area. Or perhaps all this outrage at crustacean deaths, whose presence in proximity to the works was just a coincidence, was a rage against nature and what she does and we cannot fathom. Or it was the ravings of demented people, their minds poisoned against growth, their brains addled by degrowth rhetoric and their willingness to condemn a region to unemployment. Priority for state and capital appears to be development at any cost, as long as some chips of money crack off for someone, somewhere. When there are big profits to be made, the past is but a joke. To care about all that has taken place behind closed doors, invisibly in the air, in the water, in the mud, is to submit thinking to the toxic effects of a conspiracy. No history can possibly ever be told in its entirety. Why bother? And so it is, apparently, better not to know, but just to hope, again, that the future will be better than the past.

An independent review of the workings of the South Tees Development Corporation, the body overseeing the Teesworks development, led by Ben Houchen, was ordered by the Levelling Up secretary of the reigning government in May 2023. The panel undertaking the review was to be appointed by him. Under investigation are aspects such as the mechanism for distribution of the contract for the site to two local developers, without a process of public tender, as well as claims about their acquisition of at least £45 m in dividends from the project in three years, with no evidence of any investment in it. In short, it became necessary, in the Minster's words, to investigate 'serious allegations of corruption, wrongdoing and illegality'. On the day of this announcement, it became known that the Freeport director was leaving his role to move back into the private sector.

Notes

1. Jon Warren, *Industrial Teesside, Lives and Legacies*, Palgrave Macmillan, Basingstoke, 2018, p. 161.
2. Michael Wolff, 'How I Designed the Labour Rose', *Design Week*, 19 November 2019: https://www.designweek.co.uk/issues/18-24-november-2019/labour-rose-design/.
3. *The Roundel*, 5/1991, p. 114.
4. *The Roundel*, 5/1988, p. 98.
5. *The Roundel*, 5/1988, p. 114.
6. *The Roundel*, 2/1989, inside cover.
7. *The Roundel*, 2/1989, pp. 38–9.
8. *The Roundel*, 5/1988, p. 124.
9. *The Roundel*, 3/1990, p. 52, pp. 60–4.

10. R.D. Semba, 'The Rise and Fall of Protein Malnutrition in Global Health', *Annals of Nutrition and Metabolism*, Vol. 69 (2) (2016): 79–88.
11. *The Roundel*, 3/1990, p. 61.
12. *The Roundel*, 6/1988, pp. 148–51.
13. *The Roundel*, 5/1992, pp. 98–99.
14. ICI Magazine, December 1957, p. 398.
15. All quotations relate to Charles Spence, 'The ICI Report on the Secrets of the Senses: In Association with Oxford University', 2003.
16. The ingredient of Quorn was no longer produced by ICI. In 2003, AstraZeneca sold Marlow Foods to Montagu Private Equity, who sold it on to Premier Foods in 2005. Quorn began to boom, due the expansion of culinary vegetarianism and veganism. Indeed from that plant comes the much-heralded Quorn for the vegan steak bake, sold at the bakery chain Greggs. It was sold again in 2011 and, then again, in 2015, to Monde Nissin Corporation, a firm based in the Philippines.
17. Aldous Huxley, *Brave New World: A Novel*, Harper and Row, New York, 1946, p. 59.
18. Aldous Huxley, *Brave New World: A Novel*, Harper and Row, New York, 1946, pp. 198–9.
19. Aldous Huxley, *Brave New World: A Novel*, Harper and Row, New York, 1946, pp. 237–8.
20. See, 'Clothes That Love You—Textiles to Touch the Inner You', *Pigment & Resin Technology*, Vol. 32 (4) (2003): 266.
21. The words were included in an ICI press release.
22. James Wilson and Chris Tighe, 'ICI Spin-Offs Remain Successful', *Financial Times*, 24 June 2007.
23. Tracy Shildrick, Robert MacDonald, Colin Webster, and Kayleigh Garthwaite, *Poverty and Insecurity: Life in Low-Pay, No-Pay Britain*, Policy Press, Bristol, 2012, p. 127.
24. See Charles Spence, 'Olfactory Dining: Designing for Dominant Sense', *Flavour*, issue 4, article 21, 2015, https://flavourjournal.biomedcentral.com/articles/10.1186/s13411-015-0042-0.
25. Toni Guillot, 'Wife Blames Terminal Cancer on Washing Husband's Dusty Work Overalls for Decades', *Teesside Live*, 12 April 2022: https://www.gazettelive.co.uk/news/teesside-news/wife-blames-terminal-cancer-washing-23657257.
26. David Huntley, 'I Was Virtually Eating Asbestos at Times' Says Retired Vicar Who Worked at ICI', *Teesside Live*, 20 June 2018: https://www.gazettelive.co.uk/news/teesside-news/i-virtually-eating-asbestos-times-14806747.
27. This is explored in Mareike Bernien and Kerstin Schroedinger's 2014 film *Rainbow's Gravity*.
28. Anni Albers, *On Weaving*, Studio Vista, London, 1965, p. 15.

29. https://www.flickr.com/photos/nekoglyph/23818756218/.
30. See: https://www.rspb.org.uk/reserves-and-events/reserves-a-z/saltholme/.
31. https://www.gazettelive.co.uk/news/teesside-news/you-see-it-crazy-cloud-23711592.
32. All clouds named to date are illustrated and described in *International Cloud Atlas: Manual on the Observation of Clouds and Other Meteors* (WMO-No. 407): https://cloudatlas.wmo.int/en/home.html.
33. Ernst Bloch, 'Better Castles in the Sky at the Country Fair and Circus, in Fairy Tales and Colportage', *The Utopian Function of Art and Literature* (translated by Jack Zipes and Frank Mecklenburg), MIT, Cambridge, MA, 1988, pp. 174–5.
34. Ernst Bloch, 'Better Castles in the Sky at the Country Fair and Circus, in Fairy Tales and Colportage', *The Utopian Function of Art and Literature* (translated by Jack Zipes and Frank Mecklenburg), MIT, Cambridge, MA, 1988, p. 174.
35. J. Ellard Gore, *Star Groups: A Student's Guide to the Constellations*, Crosby, Lockwood and Son, London, 1891.
36. Shivali Best, 'Twelve New Constellations Spotted Over UK—Including "Parmo" Above Middlesbrough', *The Mirror*, 19 November 2019: https://www.mirror.co.uk/science/twelve-new-constellations-spotted-over-20912862.
37. Owen Hatherley, *A New Kind of Bleak: Journeys Through Urban Britain*, Verso, 2012, pp. 51–2.
38. Owen Hatherley, *A New Kind of Bleak: Journeys Through Urban Britain*, Verso, 2012, p. 38.
39. https://www.gazettelive.co.uk/news/teesside-news/watch-redcars-iconic-coke-ovens-24343198.
40. https://www.middlesbrough.gov.uk/elections/election-results/2017-tees-valley-mayoral-election.
41. Lee McGowan, *Preparing for Brexit: Actors, Negotiations and Consequences*, Palgrave Macmillan/Springer International, 2017, p. 22.
42. See Dominic Webb and Ilze Jozepa, *Government Policy on Freeports*, Research Briefing, 22 February 2023, House of Commons Library.
43. The magazine *Private Eye* has been one publication investigating and documenting occurrences around ownership and the Freeport.
44. South Tees Development Corporation, *South Tees Regeneration Master Plan*, November 2019, p. 85.
45. As trumpeted on the promotional website: https://eastcoastcluster.co.uk/.
46. Kelley Price, 'MGT Fire Smoke "Stinks" Say Neighbours as Water Shields are Set Up to Reduce Smells': https://www.gazettelive.co.uk/news/teesside-news/mgt-fire-smoke-stinks-say-22203102.

47. Department for Levelling Up, Housing and Communities, Policy paper: 'Levelling Up the United Kingdom: Executive Summary', 2 February 2022.
48. Chloe L. Eastabrook, Miguel Morales Maqueda, Charlotte Vagg, Joyce Idomeh, Taskeen A. Nasif-Whitestone, Poppy Lawrence, Agnieszka K. Bronowska, John H. Bothwell, Brett J. Sallach, Joe Redfern, and Gary S. Caldwell, 'Determining the Toxicity and Potential for Environmental Transport of Pyridine Using the Brown Crab Cancer pagurus (L.)' (preprint on bioRxiv; the preprint server for biology).
49. Dominic Webb, Ilze Jozepa, Government policy on freeports, Research Briefing 22 February 2023, House of Commons Library, p. 29.
50. Dominic Webb, Ilze Jozepa, Government policy on freeports, Research Briefing, 22 February 2023, House of Commons Library, p. 29.
51. Houchen's comments were widely circulated on social media. He complained to the BBC that he had been misrepresented and his other comments, around his calls for support for affected fishermen and his confidence in DEFRA, were edited out. The BBC apologised for shortcomings in the broadcast edit, resulting in a lack of context.

History as Synthesis

Abstract This chapter reflects on the method deployed as a response to the scattered nature of ICI's archives and in line with a desire to tell a panoramic history, whose model is, in some regards, Walter Benjamin's *Arcades Project*. While Benjamin's history of Paris, conceived as the birthplace of consumer capitalism, is a thick, fragmented and unfinished work, and this is a slim history which aims to be complete, there are similarities. Both are drawn by resonating and recurrent themes. Both are also attracted to a variety of voices, including those often unheard, and, as well, are drawn to waste, ephemera and traces that might be overlooked, but can yield insight into how it was and is to live amidst the conditions on the ground. Examined here too is the origin of this project in relation to an exhibition, *Chemical City*, held at Middlesbrough Institute of Modern Art in 2021–2022.

Keywords *Arcades Project* · Archives · Blueprints · *Chemical City* exhibition · Historiography · Official history · *Pandaemonium* · Saltburn · Terylene · Walter Benjamin

Blueprint History

For the purposes of recounting this history of a company, a factory, a place, it is possible to draw on so many facts, piles of documents, reams of explanations of technical processes, government records, shareholder accounts, contracts, patents and reports. But equally ways might be deployed to access hopes, dreams, anxieties, fears, myths and a realm of the imagination in relation to ICI—in fantastical or horrified projections of the future of the company or the world, nightmares and illusions about what happens within the factory, desires and disillusionment expressed in relation to work and friendship, ideological insistences and conspiracy theories, the world of advertising and all the memory work after the disappearance of the company. There is a beautiful deep blue that can be uncovered in the company archives, shining out of original cyanotype blueprints for bagging machines or compressor glands.[1] A blueprint is a technical drawing using contact printing on light sensitive sheets and shows a model that should act as a guide for making something that is expected to work (See Fig. 1). What is the blueprint for writing a history—what firm white lines of investigation stand out against a blue background? The colour of old blueprints is that of the most perfect summer sky, a cloudless one. Is it possible from a context of heavy industry and material processing to stimulate poetic and creative thinking around the ways in which our lives and the lives of those who went before us were enmeshed with synthetic chemicals and the dreams that seem to ooze and seep from them as by-products? Can a history be told poetically, working through leitmotifs and resonances, puns and coincidences, repetitions that seem to suggest that there is fate at work in the history of this production, in the nature of the region. Can something passing as objective history be distilled from peculiarities that might slip under a radar that sought only the most evidential, rational details? Can history be told less through facts and figures and what reaches the official record and more through the senses, the sights, the smells, the feelings of industry? Can many voices be brought in, woven together, threaded to make something combined and complex?

rather surprising at first sight until it was remembered that vanadium would function as an anti-catalyst, and was therefore obviously manufactured in the mouse's liver.

MILES INTO PINTS

The vibrating or spinning type of mouse yielded the richest quantity of "Compound XB 29" as the new colouring matter was provisionally called. A noted physiast was consulted at this stage, who made an X-ray spectrum study of the lettuce structure of the mouse liver and correlated the results with the wave-length of the mouse's vibra-

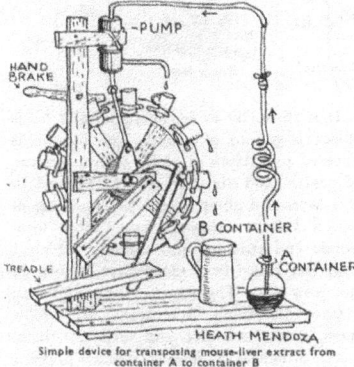

Simple device for transposing mouse-liver extract from container A to container B

tions (3456 Å). As was to be expected, the mathematical relationship took the form of the equation:

$$\eta = \frac{\lambda \pi \sqrt{-1}^x}{0.3921 w - n} + K$$

a well-known variation of the Maxwell-Lorentz transformation for converting miles per gallon into pints per head.

CONCLUSION

This of course was of purely academic interest, but our next discovery was the thing which really made the process a practical proposition. The line of reasoning was somewhat as follows: Just as we have ordinary (light) water and "heavy" water, ordinary (light) beer and "heavy" beer, so we may have ordinary (light) mice and "heavy" mice, i.e. rats. The final conclusion reached, which was triumphantly supported by experiment, was that mice themselves could be quite well dispensed with, and all that was required for the efficient working of the process was a Member of the Institute of Chemical Engineers (M.I.C.E. for short) and a few unconsidered trifles of plant, with recoveries, residues and what-not. The details will no doubt fail to interest you.

THE NEXT PROBLEM

I am already hard at work on my next problem, which is to find an economical method of extracting the blue colouring matter out of blue pills, blue-prints and blue jazz. You will no doubt be awaiting my next publication with eager anticipation.   S.W.D.

A blue-print showing device for converting melody into sound waves, which are then cut into suitable lengths for extracting blue pigment

Fig. 1   Entertainment in the February 1938 edition of the ICI Magazine

Sources

Piecing together a history of ICI is overwhelming in contradictory ways: there is too much and too little at once. There are some company histories, written at the highpoint or at the upward curve of its existence—such as the one commissioned by ICI in 1965, a two-volume study *ICI: A History* by W.J. Reader, published in 1975. Or one might consult Carol Kennedy's *ICI: The Company That Changed Our Lives*, published to mark ICI's diamond jubilee in 1986, and updated in 1993 as demergers took place. This is popular industrial history, boosterish in tone. The volumes are triumphal or celebratory in tone. Failures get few monuments. There are reams of business press reports, which follow the rise and fall of the company in real—or press—time. Traces of projections, speculations and deals are logged on legacy websites, pixelated ruined moments in a multi-decade process of ruination.

There are plentiful in-house company materials, the sober-looking journals of the 1930s and 1940s, and the magazines from the 1950s and 1960s, which revel in dramatic photography and stylish watercolour and pen illustrations. These magazines are oriented towards modernity, futurity and growth, while dispensing knowledge of the specialised jobs and processes in an industry. They fall somewhere between advertising, self-congratulation and the tracking of social and company anxieties. In a fashion akin to the middlebrow magazine *Reader's Digest*, the magazines served also to provide self-improving general knowledge, with features on 'Argentina Today', 'Runcorn: A Peep into the Past', 'Winter Pruning', 'Old Bentley versus New "Mini"'. The magazine rendered an image of the company as it wished to be known to its workers and supporters. It continued, renamed *The Roundel* in 1987, until 1993, reporting on sales forecast, mergers, demergers, plans for expansion, the challenges of recomposing global markets. There are political records, government ones and those from campaigners who monitor what happens and what is to come. There are contributions to analyses of management techniques in *Harvard Business Review* or elsewhere. There are several academic studies from sociologists exploring labour organisation—sometimes masked, if what is to be revealed exposes too much for some who are vulnerable. There are a host of photographic compilations of industrial and urban heritage and newspaper features on the past as we did or did not know it. These capture something about how it was to labour in the factories or to live amongst the installations. There are many resources of oral history, sometimes directed by professionals in museums and charities, other times swirling around self-organised on the sluices of online media. Here, arguments about memory, circulations of rumours contend with precise recall or amateur and specialist knowledge.

Should a study, such as this, wish to expand its remit beyond the data of a company that was founded, made things and closed, should it desire to understand what it conceives of as the totality of the presence of a company in a place, even over the short time of less than a century, it is hard to set limits on what might be consulted—especially if it is the case that what came before and after must also be part of what was. There is too much and it ever exceeds itself.

But it is also the case that, amidst all this plethora, there are only fragments. Even the archive of the company is fragmented. A tone of exasperation seems to exude from the archive note pertaining to the University of Manchester's archive of ICI Dyestuffs Division and predecessor companies. This archive holds the 'surviving records' of a part of the company and its predecessors. Much is missing, which appears, to the archivist, ironic:

> Although ICI was noted for its complex system of management, it does not appear to have employed modern records management procedures. Selection of records for preservation seems, where it existed, to have been by perceived historical interest, rather than an overarching view of achieving a comprehensive official record of activities.[2]

Key record series are missing. There are no minutes to be found of the various committees that were crucial for running the organisation. A bureaucratic organisation failed to apply bureaucracy to itself. The files on major projects and topics are 'fragmentary': 'Some of these files survive particularly for the 1940s, but there is almost total absence of these files for the 1960s onwards. It must be concluded that these records were not considered to be worth retaining'. It is necessary to deal with what others, unknown persons, considered worthy or not of retention. Personnel records were rarely kept, but ICI kept its own promotional materials: internal staff guides, manuals and in-house publications. The operation was vast and generated imagery, films, advertising, engineering drawings, scientific papers, reports, samples, products, packaging and so on.

John Wheeler's report 'The ICI Archive in the North East' expressed similar sentiments: 'ICI seems never to have had a clear and consistent policy of archiving for historical reasons'.[3] There was from the start a system of Central File Reports and other early documents were eventually stored there, amounting to around 36,000 surviving items which comprise 'a miscellany of company and local history'. But many things were discarded after five years, and, Wheeler notes, even 'important registries such as the Directorate did not escape wholesale destruction when the attitude of the day was to be a slim organisation, unburdened by the past'. For a while, as the company's fortunes waned, what constituted an archive was stored in the Billingham offices of Synetix, the ICI catalyst business. But Synetix was sold off and a new home had to be found. The material was sifted and some was destroyed because it gave away personal information. The amateur archivists rued some of the decisions:

> The personal data issue which did cause some distress to us from an historical point of view was that of fatal accidents. There was a collected report of a few fatal accidents which occurred in the 1920s. It gave brief details of the accident (no more than would typically appear in the press) and what the company's actions were with respect to the widow and dependents. The company considered that to be far too sensitive and it has been destroyed.

A percentage of the material was selected by the archivists for preservation and the bulk shredded. A large quantity of what remained was then concentrated at Teesside Archives—maps, plans and posters, engineering drawings, photographs, personnel and trades union records (which are still closed). There are more photographs at Beamish: The Living Museum of the North, and there are social club records and publications at Catalyst Science Discovery Centre and materials on scientific and technical processes at the Science Museum Library and Archives. Frank Ewart Smith's papers are catalogued at the Institute of Mechanical Engineers, and other scattered personal papers from technical and managerial staff can be tracked down. The Wellcome Library has one copy of *The Roundel*, as well as an exhibition catalogue, alongside plenty of papers relating to ICI pharmaceuticals. Across the road, the British Library has various issues of ICI magazines, and at least some of *The Roundel*. The multiple catalogue entries express a certain insecurity about what is there: 'Ceased in 1987?' Upon enquiry at the library, confusion reigned. A journal should not be in the Trade Literature archive. Someone was sent off to find it in the basement. Hours passed. Eventually, several people, on the case, informed me that they had found two hundred boxes of ICI materials, but were not entirely sure how it was ordered or if it was all catalogued. They would get back to me.

There is too much to survey, while there are also just splinters and tatters and it seems as if the perspective from which it could all be held together and a comprehensive story told, is missing or impossible to gain. John Wheeler observed that a contemporary fear of judgement of actions in the past by the standard of the day, combined with the transient nature of organisations and the use of electronic communications, mean that the record of today is lost: 'The archive we have been working on contains not a scrap of record of all those business divestments of the 1990s'.

To get above the clouds and look down on this and understand it in a comprehensive way is an impossible task. As you rise, the cloud layer intercedes and wipes out spots on the surface of the earth below. Approach from a different angle, perhaps something else is glimpsed. Understand what is there through what still remains—in memories or on the ground—that is one way of knowing what this was. Is it necessary to know what the company thought behind the facade of self-promotion? Would company records show that and would that be the truth of it all? Or were those boardroom discussions just more ideology—and the crucial parts of what was decided find their way into the press, translate into the leaps and dives of financial markets or mark themselves on the bodies and minds of those who carried out those decisions.

To think in this way may be far from traditional narrative in history. How to take elements from all of the possible prompts and synthesise, or combine them into a history that speaks both of a specific history or a factory in an area through a bounded period of time and also of the impact of industry in recent history, recounted as a parable of capitalist modernity, as it persists into the present, remaining unconcluded and contributing to all our futures, in that region, or in any region. Such history cannot be linear or progressive. It might have to remain fragmentary, aware of gaps or the incompleteness of any recounting. Aware too that what comes together to be understood as history is a complex whole, which is full of contradictions: those between the relations of production, of ownership, control and the forces of production, the workers or the contradiction between ideologies and actualities or those between hopes and realisations, or between locality and world, or state and capital or any other number of splits and clashes that make up the complicated picture of what comes to be. To give a sense of this density of life, the terrain to be studied might benefit from being opened up to a sense that various forces, perceptible and imperceptible, on the surface and beneath need to be excavated or revealed. Recurrent themes—such as clouds, dreams, birds, secrecy—tie together this shattered, conflictual ground, because it is a whole, even if it is multiple. This was a material world. The factory transformed and made and emanated things. It was also a fantasy of a world that was promised to come into being and to benefit all and it promises this again and again in different guises. From this perspective, the recurrence of themes evinces the extent to which history may well be, for us, in the never-ending days of capital's rule, an eternal return, a circling, an inability to shake off the past, a condemnation to repeat and repeat, having failed to learn any lessons. It can feel that way sometimes. And amongst all that can be amassed, whether it be the company-authored stories of origins or the online observations of life as a 'smoggy', there is a lot of repetition.

Method

This book is slim, but it imagines itself to match up, methodologically, in some regards, with the great, thick, unfinished 'history' of nineteenth-century Paris, Walter Benjamin's *Arcades Project*.[4] In the *Arcades Project*, written in the late 1920s and 1930s, Walter Benjamin wrestled with how to present the fullness and manifoldness of historical Paris, from its underground railways and catacombs to its modes of dress and lighting to the phantasmagoric visions held in the heads of citizens, artists, philosophers or police. He aimed to capture something of the multi-formed totality of nineteenth-century Paris. To understand Paris was to understand the form of rule of life instigated by Capitalism, a form of life riven with contradiction, with promise and disappointment, a form that incubates turbulent change and endless repetition, a form that produces its advocates, ideologues and no beneficiaries, as it produces detractors, naysayers and revolutionary challengers. Paris was *the* capital of the nineteenth century because capital as mode of production extended itself as sensibility into city planning, life and love for sale, consumerism and individualism. It was the capital, because significant history was played out here in repeated efforts to force it to deliver on its promises of liberty, fraternity and equality, as new forms of unfreedom, in wage slavery and imperial rule, military campaigns against brotherliness on a world scale and inequality, economic and otherwise persisted. Echoes of the French Revolution—a revolution on behalf of the universe—reverberated through Paris in revolutionary wave after wave. At some point, Benjamin decided on a title for his work, which had been conceived under the title *Paris Arcades*, but had exceeded that single architectural form, or rather projected the whole of the city and its history into these narrow glass and iron corridors of shops, He decided on *Paris, The Capital of the Nineteenth Century*.[5] Paris in the nineteenth century left enough fragments of thought and deeds in the archive to be comprehensively mined, in order to discover the mysteries of the mechanisms of bourgeois rule and the renewed attempts to challenge them. It was in Paris that the contradictions of bourgeois class rule were spectacularly on show. Benjamin, who spent many hours trying to find the many-faceted truth of this city, observed that 'Tens of thousands of volumes are dedicated solely to the investigation of this tiny spot on the earth's surface'.[6] How to select amongst them? His book on Paris was likely never finished. Handed down are a thousand pages of quotations, commentary. Benjamin excerpted nuggets for ideas, observations, lines of poetry or political tracts extracted from 850 books, essays and pamphlets. The materials mined spoke for themselves, it would seem. Sorted into themes, such as Fashion, Dreamhouses, Iron Construction, The Collector, The Doll, The Commune, Idleness, Mirrors, Modes of Lighting, its method was 'literary montage'.

> I needn't say anything. Merely show. I shall purloin no valuables, appropriate no ingenious formulations. But the rags, the refuse – these I will not inventory but allow, in the only way possible, to come into their own: by making use of them.[7]

What Benjamin found in the library was conceived of as the waste and refuse of history, discarded, no longer living and present thought and deeds. But in being quoted, used, or re-used, it spoke, gave off its meanings in new combinations. He acted as a ragpicker, a figure who came into his own in the period under exploration, a recycler of trash in order to sell it back to industry. In 1938, 'The Paris of the Second Empire in Baudelaire', quotes the French poet's description of a ragpicker.

> Here we have a man whose job it is to gather the day's refuse in the capital. Everything that the big city has thrown away, everything it has lost, everything it has scorned, everything it has crushed underfoot he catalogues and collects. He collates the annals of intemperance, the capharnaum of waste. He sorts things out and selects judiciously: he collects like a miser guarding a treasure, refuse which will assume the shape of useful or gratifying objects between the jaws of the goddess of Industry.[8]

Charles Baudelaire, notes Benjamin, did this as a poet, turning the dreary and workaday life of the city into a lyricism of the everyday, with all its dreams and horrors intact. Benjamin too is gathering up the debris and died-off stuff of the past, in order to enter it into a new cycle of value, one involving its historical meaning and its revolutionary potential, both of which might bear upon his present. But, there is so much material and it is pointless to construct an image of the world that is the same size and shape in every way. Selection is crucial, for it is through selection that the image of the past, always flitting by, can be captured in its truest significance. Benjamin observes:

> All historical knowledge can be represented in the image of balanced scales, one tray of which is weighted with what has been and the other with knowledge of what is present. Whereas on the first the facts assembled can never be too humble or too numerous, on the second there can be only a few heavy, massive weights.[9]

So many details on one side, the most trivial things that make up the image of the past, but it must be balanced by just a few heavy weights. On this side of the scale is the knowledge of what is present. What is it about our contemporary moment that makes sense of all the multiplicity of data from the past? What is the weight of current concerns that select judiciously from the trash heaps of the past. Those heaps of lives lived are mingled with ideologies, fantasies, dreams, things that happened and did not happen. How to pick apart this mixed and impure matter? How to refine and resynthesise it? The image of the scale evokes the pans of chemical experiment, but it is also, in as much as it must balance, a metaphor of justice, of Justitia, holding aloft her scales, weighing facts and evidence, to come to a verdict, and, ultimately, to restore balance in the world. Is the knowledge of the present making sense of the past connected to questions of justice, or laying appropriate blame in the past in relation to the ongoing harms of the present? How did we get here, and how do they work on and into now? The scattered details of the past have coagulated into immovable clumps in the present. With a full transcript, one might be ready for Judgment Day, though Benjamin reminds us, in 'On the Concept of History', that such reckoning is possible only after we have been redeemed.

> The chronicler who narrates events without distinguishing between major and minor ones acts in accord with the following truth: nothing that has ever happened should be regarded as lost to history. Of course only a redeemed mankind is granted the fullness of its past - which is to say, only for a redeemed mankind has its past become citable in all its moments. Each moment it has lived becomes a citation a l'ordre du jour. And that day is Judgment Day.[10]

Benjamin's arcades study provides a challenge to traditional historical methodologies, preferring to present the details and the tensions of developing Capitalism in Paris after the modernising schemes of Baron Haussmann and others through montage, multi-perspectivally, opening up cracks, threads, multivocal approaches and records of what was, as much as what was thought about what was. It was unfinished and what might have melded it all together can only be conjectured, on the basis of his other more complete works about Paris in that period. And perhaps, he might have left it as montage or collation, just as did Humphrey Jennings, more or less, in *Pandemonium*—though that too was a posthumous compilation and contraction of amassed materials.[11] In the context of twentieth-century industrialisation and automation, Humphrey Jennings' collected various published accounts about the coming of the machine as perceived by observers.

To approach history in these ways, prismatically and layered, presents itself as a possible model for thinking about history in relation to a chemical processing plant whose existence is shaped by its own array of operations and personnel that appear likewise, in various ways, as layered, as prismatic or multi-faceted. Could this be something like an *Arcades Project* of the Chemical Industry in the North East, organised thematically and intended to stimulate poetic thinking around the ways in which our lives and the lives of those who went before us were enmeshed with synthetic chemicals and the dreams that seems to ooze and seep from them as by-products. If Paris was the capital of high capitalism in the nineteenth century, was Teesside the centre of twentieth-century capitalism? In any case, it too waded through the trivia and serious, fleeting and monumental details of the past and tried to find their outflow into the concerns of the present, the heavy weights of environment, labour, reproduction, the home, social relations, the concerns of justice and the ongoing legacy of a mode of production in a place.

Chemical City: Origins

This study first came into being as a commission from Middlesbrough Institute of Modern Art. An exhibition, *Chemical City*, was in the planning and I was asked to be a consultant, undertaking research for the conceptual direction of the exhibition, which was to include a combination of company materials, relevant memorabilia, photographs, films and commissioned artworks. I was asked because, in 2005, I published *Synthetic Worlds: Nature, Art and the Chemical Industry*, which told the history of German chemical conglomerate IG Farben.[12] It did more than that, for it used the occasion of the invention of artificial dyes and synthetic substances as a way of telling a history of Germany from the nineteenth century into the Third Reich. It inquired into the implications of an industry developed of synthetic dyes to compensate for the fact that Germany had managed to capture minimal colonies—and therefore resources and markets—compared to other European nations and so invented substitutes in laboratories and factories. German science and technology were very successful at making *Ersatz* products, for example dyes, and it led to a massive industry. What was produced was paraded as so much better than nature's own outputs. Dyes were lightfast, waterfast, brighter and enduring. That was the promise of a rainbow world in the nineteenth century, facilitated by technology and science from companies such as Hoechst, BASF and Bayer. In the twentieth century, science and technology from the combined companies, now named IG Farben, pivoted towards the needs of war, from 1914, and, towards a synthetic life with plastics, pills, artificial fertilisers and rockets, which came to be facilitated by slave labour and concentration camps of the Nazi regime. The history moved from the production of synthetic dyes to the production of death through labour. This deadliness was a finality that appeared, perversely enough, under the sign of nature, as articulated in the Nazi rhetoric of blood and soil, the naturalness of racial divisions and the fatedness of Hitlerian rule.

Again, through the development of the chemical industry, the history I wanted to capture was panoramic. How did the emergence of synthetic colour come into being in the context of colonialism, industrialisation and Nazism, and also how did it affect the development of the senses? How could that history marked on the body and in cultural forms be told? It seemed intriguing to explore part of this same period of history, another part of this industry through what happened in Teesside.

Chemical City, the exhibition which was at Middlesbrough Institute of Modern Art, from late 2021 into the Spring of 2022, was composed of documentary materials, photographs, films, substances, as well as artworks and the products of sustainable materials research. It collated many ways of seeing and processing the industry and the legacy of what had so affected the region. Its first room gathered together the material culture of the company, brochures and magazines, posters, dye labels from the 1920s and 1930s, photographs of men at work and at company-provided leisure, a pink Alkathene cup and saucer, company ties, banners with the ICI logo, safety gear and brightly patterned artificial silk dresses, an ICI film about chemical processing, the last Perspex sheet made at Wilton on 15 January 1970.

Above these archive chips, utilising the high ceiling of the gallery was *Future Booster* (2021), a sculpture by Annie O'Donnell, teal and safety yellow with blue netting, reminiscent of gantries and industrial installations that stripped and snake through the landscape. In the next room, O'Donnell had created a red metal scaffolding with hanging fabrics, like safety wear and overalls. On the wall were collages, *Synthetic Catalysts*, incorporating family photographs, traces of love led in this landscape, bodies fed by the fruits of labour in the chemical factories, homes furnished by the products, views from windows overshadowed by the plant. If these dug into the memories of past existences, on adjoining walls were diggings into the ground around the region. The theme was the Billingham underworld, and artefacts and documentation referred to what was mined, to the substances that subtend Billingham, its rock salt and gypsum selenite, marl, all recovered by for processing. Here, in this room, was work by Onya McCausland. A pyramid of household paints made by the artist was framed by *Saltburn 54°34 07.37 N 0°57 42.87 W* (2021), a 13-metre-long wall painting, in an earthy browny yellow, named 'Saltburn Ochre', paint made by the artist from waste ochre at a mine water treatment site in Saltburn in the Tees Valley, The colour was recycled from the spill offs of a disused ironstone mine, which had stopped producing in the 1960s, and had been polluting waterways, until campaigners had pressed for a water treatment scheme in 2015. The pigment was combined with an environmentally responsible potassium silicate or water glass binder. This ochre joined the other ochres made from other sites, each one specific, each one different, 'formed from the specific environmental, ecological, geographic idiosyncrasies of that place only'.[13]

In the next room were more traces of the reach of ICI, a reel of promotional films tracking across the era of the company's existence, its hopes, promises and demise, and a map from 1929, depicting its global reach. The Empire in the name Imperial Chemical Industries is made explicit. The captions in the map's corners declare that 'The Company Plays a Part in the Safety and Progress of the Empire, not only as the basis of its leading industries, but as a factor in the development of its natural resources'. For development, read extraction and exploitation. The rest of the room was taken up by a trade fair style display of more sustainable fibres and fabrics made in response to the ecological impact of synthetics, curated by Lynne Hugill with examples from students at Teesside University and international start-ups: trainers made from coffee waste materials, biodegradable fabrics, such as banana viscose or vegan leather made from cactus, ecological processes for making denim.

The film, *Birds of Teesmouth* from the RSPB Film Unit in 1966, played on a gallery screen. This was the presentation of a more or less accord between nature and industry, a bird life lived between estuary mud and factory chimneys. Was it an anomaly or a model? Was it a version of the future or something of a problem to be solved or a mode of life that would eventually become unsustainable? Did it hold any lessons for other lives in the vicinity? The flight lines out of all this toxic mess are visible, new things to produce, less burden on the earth, new industries, or something after industry. How do the hopes translate into the really developing Charter City, the Freeport, the Teesport, the gigafactories and carbon capture off shore hulks? Nature will force itself through the cracks between security fences and perimeter walls and along the edges of containers transformed into construction site offices for the years of building and rebuilding,

A final work in the exhibition was by Katarina Zdjelar, *Not a Pillar Not a Pile (Dance for Dore Hoyer)* 2017/2021. It was a multi-layered work that took at its centre a dance studio that was set up in the ruins of Dresden just after the Second World War, its first dance drawing on

Communist artist Käthe Kollwitz. Little remains of this experimental process of choreography, which occurred in the days just before the founding of the now-extinct GDR. The choreographer was destitute and working in the one remaining intact room of Mary Wigman's former studio, the rest being bombed. Archival traces of what occurred here are solely some photographs and a partial musical score. These fragments form the basis of a multi-channel video work, in which dancers, activists and performers interpret in their own way Hoyer's *Tanz für Käthe Kollwitz*.

Another historical resonance was introduced in the work: costumes for the reconstructed dances, shown on film, were based in designs from the female workers of PAUSA, a German Jewish textile factory in Mössingen, with a Bauhaus sensibility, where the workers had revolted against Hitler's rise to power in 1933. The revolt was crushed. The firm was given to an Aryan entrepreneur. Factory and fabrics, bodies clothed, in protest, in dance, history that has happened and reconstructed. Zdjelar's work, with lines gouged out of the gallery floor, screened partial visions of twisting bodies, and it brought to the fore gaps in the archives and how imagination, or imaginative reconstruction, might fill them in or out.

A Last Chorus

This rumination on ICI compiles facts, myths, fictions, stuff from archives, catalogued ones, ones accumulated online, incidental ones, commercial traces, attempting a synthesis. It plugs the gaps of official records with other matter, other voices. It undermines or amplifies voices of experience and those of rumour by the data of record. It hopes it has found some massive heavy weights that balance the tangle of detail. But those few heavy weights might, actually, prove to be the weight of a history turned fate, the recurrence of motifs, repetition of themed in a spot, a geography evincing various versions of the same from itself again and again, until it is over. It hopes to have produced new ways of seeing and processing this layered, tangled history (See Fig. 2). It hopes to have fixed something, made what is fluid and volatile solid and usable, at least for a moment.

A poetically attuned sensibility may be alert to coincidences, resonances, themes and leitmotifs. Perhaps what catches the attention is a result of distortion: what leaps out of the archive, of the grey literature, of the photo repositories, what emerges through the selected tags in online

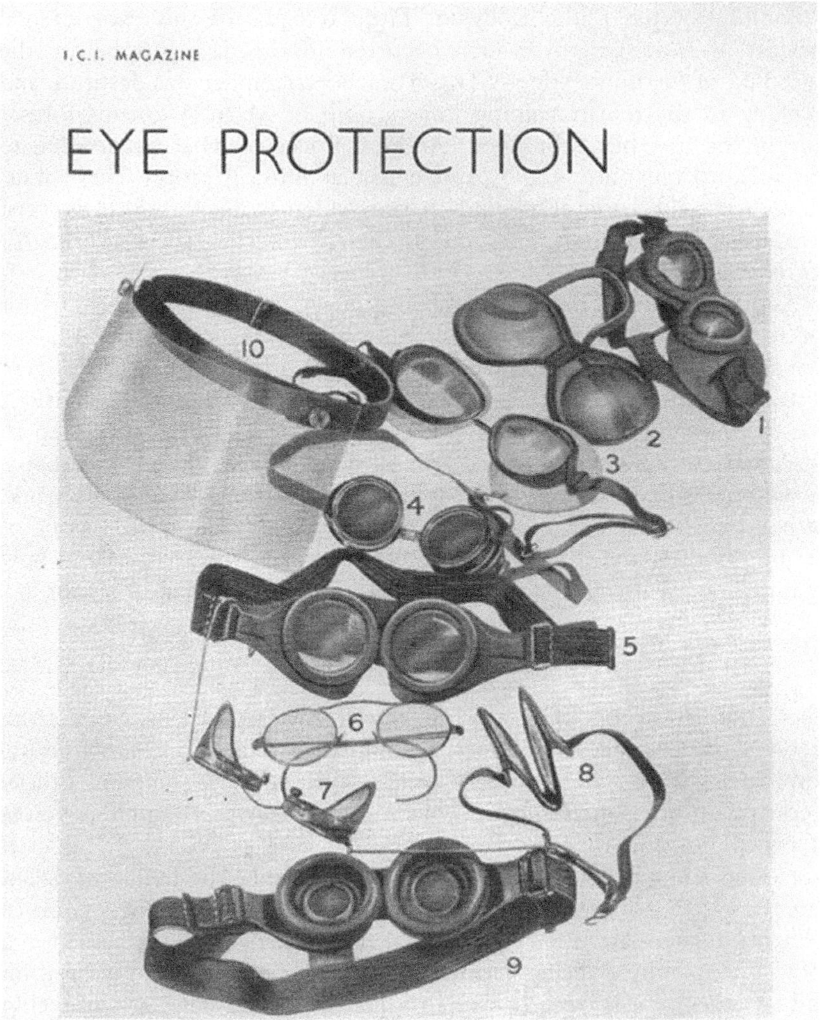

Fig. 2 Safety advice in the ICI Magazine, February 1938

searching and surfing, is what is already at work in the imagination of the researcher. Could the truth of the company ever be found? Are the few massive heavy weights the weights that weigh down the researcher's deranged mind? I found and found out only what I was looking for: persistent but persecuted birds, dreams of synthetic abundance, recurrent stinks, natural and unnatural clouds, the fate of Hercules, ground level injustice, corrupted politics amidst ruined nature, threads tangling and unravelling.

In volume three of *Modern Painters*, from 1856, John Ruskin, critic, coined the term 'pathetic fallacy'.[14] It was directed at and against overly sentimental poetry, and what he perceived as an emotional falseness in the writings of Blake, Shelley, Wordsworth and Keats, amongst others. In poems beset by pathetic fallacy, as Wordsworth put it, objects 'come to derive their influence not from properties inherent in them', but rather 'from such as are bestowed upon them by the minds of those who are conversant with or affected by these objects'.[15] Ruskin examined the absence of truth or factuality in phrases that describe foam as cruel and crawling or the crocus as gold and spendthrift or the white rose as weeping. Such descriptions depart from the truth of the object described to convey instead the strength of emotion of the describer. But this technique, he states, might grasp a truth, if used exquisitely. He quotes a couplet, concerning a poet who wishes his body be cast into the sea:

> Whose changing mound, and foam that passed away,
> Might mock the eyes that questioned where I lay.

The sea's crests described as mounds is true, is something like a scientific fact, because it conveys the heavy darkness of a wave and this wave changes in a way that is closer to the rolling of hills than to something rising and falling, an error of perception and understanding.

> Most people think of waves as rising and falling. But if they look at the sea carefully, they will perceive that the waves do not rise and fall. They change. Change both place and form, but they do not fall; one wave goes on, and on, and still on; now lower, now higher, now tossing its mane like a horse, now building itself together like a wall, now shaking, now steady, but still the same wave, till at last it seems struck by something, and changes, one knows not how, becomes another wave.[16]

The poet presents the sea as a constant flux, ever-changing, with dissipating foam, but it is also its opposite, a solid mound, heavy state that does not pass away but stays, persists, as do too the white stones, perhaps of graves, with writing on them, and gesturing to endless time and death. The sea with its ever-changing form and its permanence mocks those who look on and no longer know where the corpse resides. Is this life now calmly in its grave? Does it still despair?

Did ICI rise and fall? Was its fate anything like the schematic waves depicted on the first versions of its roundel, borrowed from Nobel Industries? Some peaking, some troughs. Or was it one wave going on, and on, and still on; now lower, now higher, now tossing its mane like a horse, now building itself together like a wall, now shaking, now steady, but still the same wave, till at last it seems struck by something, and changes, one knows not how, becomes another wave? A wave that rolls on, in other forms, with other arrangements, but still that heavy wave of heavy industry, of dumping grounds and new promises and everything suggested in the mind that is opposite to all that. Or was it like the later logo, two flattish lines ploughing through the century, looking less like a wave with its ebbs and flows and more like a cutting into the ground unremittingly, more like shafts through a landscape, up underground tunnels, roads to facilitate both transportation and the promise of fossil-fuelled modernity, a path that is an inescapable fate, an undeviating mission?

In the ICI magazine of the 1960s, the blocky colours of charts, the saturated hues of photography, the vibrant diagrams of scientific process and trade routes illustrated a hopefulness, a sense of eternal sunshine that came to seem distant. What made it distant? That the factories no longer produced jobs for life and that meant a disappearance of that intangible commodity, if it was one, whose name was never far away, when talking of skilled manual labour of the recent past: pride in labour, dignity in work. It became distant too because it seemed that many of those synthetic products were responsible for the toxicities that harmed our planet and caused extinction and damage to animals and humans and plant life. We woke up to this damage, the damage of plastics everywhere and never ever decaying. But we also understood that there might be a future for synthetics, that there was ingenuity in making things from other things, but it needed to be done responsibly and it might have to be done if there is nothing left of the original stuff. To make anything and everything by our own hands, by tricking nature, or outwitting it, remains also part of the image of utopia.

Social, biological, chemical life coexisted, co-developed, shaped and unshaped, formed and deformed and not as an ongoing chain of progress, but as a cycle, as repeated returns to the conditions that exist, to what can be done in the circumstances, given these elements, these people, but, always also, given these policies, this demand. There were flows here, still are, flows of chemicals, of money, of bodies, or materials, of traffic. There are visibly fewer of some of those things now. Ways of life unravel, but not fully. Like Terylene threads, the process of decomposition is slow and complex, productive and destructive. Remnants remain. That they remain can sometimes be a problem or a new beginning.

Notes

1. There are examples in Frank Ewart Smith's archive at the Institute of Mechanical Engineering amongst the technical papers.
2. Note online for ICI Dyestuffs Division and predecessor companies archive. University of Manchester Library: Reference GB 133 ICI.
3. Held online: http://public.bacs.daisy.websds.net/PDFFiles/Articles/89065.pdf
4. Walter Benjamin, *The Arcades Project* (translated by Howard Eiland and Kevin McLaughlin), Belknap Press, HUP, Cambridge, MA, 1999.
5. Walter Benjamin, *The Arcades Project* (translated by Howard Eiland and Kevin McLaughlin), Belknap Press, HUP, Cambridge, MA, 1999, p. ix.
6. Walter Benjamin, *The Arcades Project* (translated by Howard Eiland and Kevin McLaughlin), Belknap Press, HUP, Cambridge, MA, 1999, pp. 82-3.
7. Walter Benjamin, *The Arcades Project* (translated by Howard Eiland and Kevin McLaughlin), Belknap Press, HUP, Cambridge, MA, 1999, p. 460.
8. Walter Benjamin, *Selected Writings: 1938-1940* (translated by Howard Eiland, Gary Smith, Rodney Livingstone), Belknap Press, HUP, Cambridge, MA, 1996, p. 48.
9. Walter Benjamin, *The Arcades Project* (translated by Howard Eiland and Kevin McLaughlin), Belknap Press, HUP, Cambridge, MA, 1999, p. 268.
10. Walter Benjamin, *Selected Writings: 1938-1940* (translated by Howard Eiland, Gary Smith, Rodney Livingstone), Belknap Press, HUP, Cambridge, MA, 1996, p. 390.
11. Humphrey Jennings, *Pandaemonium 1660-1886: The Coming of the Machine as Seen by Contemporary Observers*, Andre Deutsch, London, 1985.

12. Esther Leslie, *Synthetic Worlds: Nature, Art and the Chemical Industry*, Reaktion, London, 2005.
13. *Then and Now*, online booklet from MIMA, p. 71: https://issuu.com/mimagallery/docs/chemical_city_then_now_final_tp_smaller_
14. See John Ruskin, Of the Pathetic Fallacy, *Modern Painters*, Vol. iii, pt. 4, John Wiley, New York, 1885.
15. William Wordsworth, *The Poetical Works of William Wordsworth*, Vol. 4. W Paterson, Edinburgh, 1883, p. 199.
16. John Ruskin, 'Of the Pathetic Fallacy', *Modern Painters*, Vol. iii, pt. 4, John Wiley, New York, 1885, pp. 160–1.

Index

A
Accidents, 8, 10, 15, 16, 21, 68, 72, 78, 91, 142
Acetic acid, 2
Acetylene, 2
Acrylic, 3, 48, 81, 101
Aerosols, 50
Africa, 3
African migratory locust, 30
Agfa, 22
Air freshener, 102, 105, 108
Air Intelligence, 27
Alamy Stock, 119
Albers, Anni, 113, 134
Alkali Act 1863, 78
Alkalis, 13, 14, 18, 72
Alkathene, 37, 38, 41, 49, 74, 149
Aluminium sulphate, 2
Amber, 3, 4
Ammonia, 2, 19, 21, 22, 44, 46, 62, 73, 77, 99
Ammonia Avenue, 23, 24
Ammonia Group, 15
Ammonium nitrate, 19, 26, 44

Angel, Ron, 86, 91
Anhydride, 6, 21
Apprentices, 70, 71
Arachin, 32
Arcades Project, 144, 147, 155
Ardil, 32, 37, 39, 40
Asbestos, 111, 112, 114
Asperitas, 120
Astra AB, 101
AstraZeneca, 101, 119, 134
Atacama Desert, 18
Atomic bomb, 50
'Atomic Energy Disclosures', 29
Australia, 9, 18, 101, 129

B
Bakelite, 4, 23
Bananas, 2
Barnett, Clifford, 81
BASF, 22, 101, 148
Baudelaire, Charles, 146
Bauhaus, 114, 151
Bayer, 22, 148
Beeswax, 3

© The Author(s), under exclusive license to Springer Nature Switzerland AG 2023
E. Leslie, *The Rise and Fall of Imperial Chemical Industries*,
https://doi.org/10.1007/978-3-031-37432-6

158   INDEX

Benjamin, Walter, 22, 23, 57,
    144–147, 155
Beynon, Huw, 58, 85, 86, 91, 92
Bicarbonate of soda, 13, 14
Billingham, 6, 16, 18, 21, 23, 26, 27,
    29, 33–35, 37, 44–50, 56, 57,
    67, 73, 76, 78, 81, 85, 88, 97,
    99, 105, 111, 112, 114, 119,
    129, 149
Bioethanol, 124
Biofuel, 124
Biomass, 99, 100, 124, 130
Biopol, 99
Birds, 8, 18, 73, 75, 76, 82, 119,
    120, 122, 131, 144, 153
Birds of Teesmouth, 75, 150
Blade Runner, 115
Blair, Tony, 110
Blake, William, 153
Bloch, Ernst, 121, 135
Blueprints, 138
Brazil, 49, 97
Brecht, Bertolt, 62–64, 89
Bri-Nylon, 118
British Committee on Industry and
    Trade, 10
British Dyestuffs Corporation Ltd, 12
British Empire, 32
British Malaya, 3
British Petroleum, 95
Brunner, Mond and Company Ltd, 12
Brussels, 129
Butadiene, 2, 37, 49

C
Carbon bisulphate, 13
Carbon capture utilisation and storage
    technology, 130
Carbon disulphide, 10
Cart horse, 8
Casein, 4

Catalin, 4
Caustic soda, 10, 13, 37, 119
Cavum, 121
Celluloid, 4
Cellulose, 3, 4
Central Agricultural Control, 48
Ceramic, 3
Chardonnay silk, 10
ChemCo, 85, 87, 88
ChemicalCity exhibition, 148, 149
Chemical Workers Song, 86
Chemische Fabrik
    Griesheim-Elektron, 22
Chemische Fabrik vorm. Weiler Ter
    Meer, 22
Chemistry at your Service exhibition,
    29, 31, 35
Chile, 18, 19
China, 18
Cicero, Marcus Tullius, 107
Cirrus fibratus homomutatus, 122
Cirrus floccus homomutatus, 122
Cleveland blast furnace, 9
Cleveland Cine Club, 54
Clouds, 8, 33, 44–46, 48, 49, 62, 65,
    73, 79, 80, 86, 109–111,
    118–122, 125, 135, 143, 144,
    153
Coal, 2, 6, 7, 9, 14, 22, 34, 53, 76,
    81, 88, 125
Cockaigne, 48
Colour Advisory Department, 51
Copal, 3
Cosak, 118
Cotton, 3, 10, 11
Cowper Bewley, 129
Crex, 13
Crimplene, 40, 41, 118
Crookes, William, 19, 99
Croydon Mouldrite, 23
Crustaceans, 131
Cyanide, 18, 86

Cyclohexane gas, 86

D
Daily Express, 36
Darlington, 7, 45, 55, 77, 131
Darvic, 41
Dengue, 30
Diazotrophs, 11
Dichloroethylene sulphide, 22
Dingle, Charles V., 65
Divestments, 95, 97, 100, 108, 109, 143
Dorman Museum, 117
Dreyfus brothers, 4
Drikold, 15, 44–46
Drugs, 30, 31, 52, 97, 101, 105, 108
Du-Lite, 37
Dulux, 37, 51, 102, 109
DuPont, 49, 87, 99, 101

E
East Coast Cluster of Teesside and Humber, 130
Egypt, 18
Electricity, 11, 19, 75
Endurance Reservoir, 130
Enron, 101
Eston Council, 35
Eston Hills, 7
Ethylene, 37
European Alpine Goat's horn, 3

F
Factories, 2, 4, 6, 8, 10, 11, 15, 20, 21, 25, 26, 29, 35, 39, 41, 44, 47–49, 54, 62, 64, 67, 70–73, 75, 76, 79, 81, 82, 84–88, 106, 108, 111, 112, 114, 118, 119, 125, 127, 130–132, 138, 140, 143, 144, 148–151, 154

Fertilizer, 44, 96
Fibres Division, 40
First World War, 20, 22, 65
Fish smell, 77, 78, 119
Fisons, 95
Fleck, Alexander, 36
Fluctus, 121
Flux, Mike, 98
Formaldehyde, 4
Formica laminate, 4
France, 18, 43
Frankincense, 3
Freeport, 127, 128, 132, 133, 135, 136, 150
Freeth, Francis Arthur, 20, 56
French Revolution, 145
Fujifilm Diosynth Biotechnologies, 119

G
Gammexane, 30
German offensive of March 1918, 20
Germany, 10, 18–21, 27, 43, 52, 94, 112, 114, 148
Gigafactories, 129, 150
Gladstone, William Ewart, 9, 79
Glass, Ruth, 65, 66, 90
Glucose, 2
Goblin, 64
Gove, Michael, 130, 131
Gramoxone, 76
Greenwood, Walter, 11, 56
Greggs, 134
GrowHow, 118
Guano, 18, 75
Gutta-percha, 3
Gypsum, 46, 86, 149

H
Haber-Bosch process, 19
Haber, Fritz, 22

Harrogate, 40
Harvey-Jones, John, 94, 95
Harwell piles, 29
Hatherley, Owen, 122, 135
Haverton Hill, 81, 82
Head, Sir George, 7
Heavy Organic Chemical Division, 49
Heliades, 3
Hercules, 63, 64, 79, 80, 122, 153
Heroin, 52
Heysham Aviation Fuel Works, 26
Hodgson, Maurice, 50
Hoechst, 22, 148
Holoplast, 33
Hoover, 64
Houchen, Ben, 125, 133, 136
House of Commons, 77
HQ Flying Training Command, 46
Hugill, Lynne, 150
Huntsman, 101
Hussain, Khadim, 83, 84
Huxley, Aldous, 105, 115, 134

I
ICI Agrochemicals, 97
ICI Alkali Group, 12
ICI Asean Group, 98
ICI Billingham Film Unit, 44
ICI House, 50
ICI Lackenby Nylon plant, 118
ICI Method for Job Appraisement for General Worker Jobs, 67
ICI Report on the Secrets of the Senses, 102, 134
ICI roundel, 36, 95
I.C.I. Weekly Staff Agreement, 87
If I film at 2fps, 54
IG Farben, 22, 23, 94, 112, 148
India, 9, 18
Indigo, 2
Industrial revolution, 3, 78

Ineos Acrylics, 102
Insecticides, 30, 50
Instrument of Consent, 129
International Conference on Genetics, 14
Ironopolis, 9
Ironstone mining, 7

J
Jennings, Humphrey, 147

K
KayNitro, 43
Keats, John, 153
Kirkleatham, 129
Kneeling Giant, 122
Knowledge economy, 124
Korea, 34

L
Labour power, 6
Labour's Plan for Science, 118
Laissez-faire capitalism, 7
Land of Milk and Honey, 48
Lang, Fritz, 86
Leathercloth Division, 37
Leicester, 34
Levelling Up, 130, 131, 133, 136
Liebig, Justus von, 18, 19
Lighting, 47, 70, 105, 144
Love on the Dole, 11
Luilekkerland, 48
Lutyens, W.F., 12–15
Lynmouth, 46

M
MacLeod, Norman, 51
Malaria, 30
Marlow Foods, 99, 101, 134

Mass Observation, 64
McCausland, Onya, 149
Medhurst, Franklin, 79, 82, 91
Medical officers, 10
Melamine, 3, 48
Metals Division Publicity Department, 70
Methanex, 102
Methanol, 2, 95, 99, 102
Metropolis, 80
Middlesbrough, 6, 7, 9, 10, 54–56, 65, 66, 79, 80, 83, 84, 114–117, 122, 131
Middlesbrough High School, 64
Middlesbrough Institute of Modern Art, 148, 149
Milk, 4, 13
Milky Galalith, 4
Ministry of Munitions, 20
Mohindra, Shibani, 108
Molecular gastronomy, 110
Molecules, 2, 6, 31, 103
Monde Nissin Corporation, 134
MONIAC (Monetary National Income Analogue Computer), 33
Monkey nut, 32
Monroe, James, 75
Mould, 3, 32, 51, 99
Moulton, John Fletcher, 20
Muriate, 13
Mustard gas, 22
Myrrh, 3

N
Naphtha, 81, 130, 131
Napoleonic Wars, 18
Nationalisation, 36
National Starch, 102
Nature, 2, 3, 6, 8–10, 43–46, 53, 54, 62, 70, 72, 73, 75–77, 79, 101, 106–108, 119–122, 125, 138, 143, 148, 150, 153, 155

Nazi regime, 148
New feudalism, 83
New Marske, 84, 120
New Working Arrangement, 86, 87
Nichols, Theo, 85, 86, 91, 92
Nineteen eighty-four (The Last Man in Europe), 115
Nitro-Chalk, 22, 43, 44, 46
Nitrogen, 11, 18–23, 44, 62, 95, 110
Nitrous oxide, 53
Nobel Industries, 154
Nobel Industries Ltd, 12
North Sea, 7, 46, 79, 125, 130
Novavax, 119
Nylon, 3, 34, 37, 40, 41, 44, 48, 52, 71, 72, 101

O
O'Donnell, Annie, 149
Odur, 3
Oil, 2, 14, 22, 34, 37, 43, 53, 75, 119
Olefin products, 37
O'Neill, Brendan, 102
ONE NorthEast, 110
Open cast strip mining, 7
Operation Cumulus, 46
Oppau, 20, 21
Orwell, George, 25, 57, 63, 64, 82, 89, 115
ORWO, 112
Outline Plan for the North East Development Area (1949), 35

P
Pakistan, 33
Pandemonium, 147
Paraquat, 76
Paris, 14, 144, 145, 147
Parkesine, 3
Parmo and Chips, 122, 124

PAUSA, 151
Pearl Dust, 13
Peenemunde, 27
Penicillin, 32
Perspex, 31, 33, 34, 37, 41, 81, 149
Peru, 18
Petroleum, 3, 43, 99, 122
Pharmaceuticals, 6, 23, 100, 101, 142
Phillips, William, 33
Photography, 4, 5, 15, 118, 140, 154
PIAT (projector, infantry, anti-tank), 27
Pine resin, 3
Pitch Lake of La Brea, 3
Plantation, 3, 10
Plastics, 2–4, 6, 23, 30, 31, 37, 39, 43, 44, 48, 49, 52–54, 62, 95, 101, 148, 154
Plastics Division, 23, 35, 81
Plastiglomerates, 53
Poland, 31
Pollution, 34, 53, 65, 66, 77, 79–83, 88, 107, 122
Polyethylene, 25, 37, 41, 95
Polyethylene terephthalate, 34
Polypropylene, 37, 40, 48, 101, 109
Polythene, 31, 37, 52, 71, 85, 100
Polyvinyl chloride (PVC), 37, 41, 81, 95
Port Darlington, 7
Porton Down, 20
Press-Button Age, 50, 102
Propathene, 48
Pruteen, 99
Ptolemy, 122
Pyridine, 131, 132, 136
Pyroplastic, 53

Q

Queen Mary, 4
Quest, 2, 102, 104
Quorn, 99, 100, 119, 134

R

Race, Ernest, 33
Radar, 25, 31, 138
RAF Middleton St George, 45
Raik, Etienne, 41
Railway, 7, 9, 14, 25, 127, 144
Ramsay Laboratories, 19
Rank Hovis McDougall, 99
Rank, J. Arthur, 99
Rayon, 10
Reactions, 2, 3, 11, 26, 37, 43, 44, 67, 95, 112
Redcar, 55, 83, 125, 126, 128, 130, 131
River Tees, 7, 19, 83
RMS Mauretania, 52
RNA, 100
Rodgers, Bill, 77, 78, 91
Rowntree, Griselda, 65
Royal Air Force, 45, 65
Royal Dutch-Shell, 26
Rubber, 2, 3, 10, 37, 52
Ruskin, John, 153, 156
Russia, 9

S

Salford, 11
Saltburn, 84, 149
Saltholme, 119, 120
Schlaraffenland, 48
School, 33, 35, 64, 65, 68, 70, 71, 78, 81, 83, 117, 122
Scott, Ridley, 115, 116
Seal Sands, 75, 129
Seaton Carew, 75
Second nature, 106, 107
Sensism, 103, 104, 107, 108, 110, 111
Sesqui, 13

Sheffield, 25
Shelley, Percy Bysshe, 153
Sheppard, Raymond, 73
Shift system, 85
Ship, 121
Shipbuilding, 7, 9, 16
Shopping centres, 40, 81, 110
Silicate of soda, 13, 14
Silk, 3, 11, 13, 149
Singapore, 49
Slavery, 10, 144
Smith, Frank Ewart, 21, 26, 27, 56, 70, 71, 90, 142, 155
Soda ash, 13, 14, 37, 100
Soda crystals, 13, 37
Soma, 105
South Tees Development Corporation, 133
Soviet Union, 71
Spence, Charles, 102, 104, 110, 134
Steelmaking, 7
Stephenson, George, 7
Stockton-on-Tees, 6, 9, 55, 77–79, 82, 100, 115
Strawberry yoghurt, 104
Suez Canal, 37
Sulphate Group, 15
Sweden, 27
Synetix catalyst business was sold, 102
Syngas, 2
Synthesis, 2, 19, 21, 22, 72, 112, 119, 151
Synthetic Ammonia and Nitrates Ltd, 20

T
Tallow, 3
Tanganyika region, 39
Tar, 3, 127
Tea Centre, 29
Teesplan, 82

Teesport, 80, 127, 150
Teesside Development Corporation, 110
Tees-Side Industrial Development Board, 35
Tees valley, 6, 16, 55, 79, 82, 125, 149
Tees Valley Regeneration, 110
Tees Valley Unlimited, 110
Teesworks, 127, 128, 133
Terra Industries, 101
Terylene, 34, 37, 39–41, 53, 71, 72, 118, 155
Thatcher, Margaret, 109
The Recruitment, Training and Promotion of Technical Staff, 68
The Road to Wigan Pier, 25, 63
The Roundel, 50, 97–102, 140, 142
Thompsons of Prudhoe, 125
Tortoiseshell, 3, 4, 75
Town and Country planning act of 1944, 67
Toys, 48, 52
Trimpell Ltd, 26
Trump, Donald, 120
Tsetse fly, 30
Two Black Crows: A Billingham Tragedy, 16
Tyneside, 16

U
Ulstron, 40
Unilever, 101, 102
Union Carbide, 101
United Alkali Company, 12
United Arab Emirates, 95
United States (USA), 10, 13, 97, 98, 110
Uranium, 29
Urea, 2, 3, 23, 32, 44, 95

V

Viscose, 3, 10, 13, 105, 150
Visqueen, 97, 100
Vymura, 51
Vynair, 41, 42
Vynide, 37

W

WalFlair, 51
Walker, Martin, 83
Weather Experiments, 45
Welwyn Garden City, 49
Westminster typeface, 51
Wheeler, John, 142, 143
Whitby, George, 27, 56
Whole Earth Catalog, 52
Wilson, Harold, 118
Wilton, 6, 16, 35, 37, 39, 43, 49, 71, 72, 80, 83, 85, 86, 113, 118, 122, 129, 149
Wilton International Plug & Play Chemicals Park, 129
Wind farms, 125, 126
Wolfen, 112
Wolff Olins, 95
Wordsworth, William, 153, 156
World Meteorological Organisation, 120

X

Xylonite, 4
Xylyl bromide, 22

Y

Yellow fever, 30

Z

Zdjelar, Katarina, 150, 151
Zeneca, 101